Wake Technical Community College
Health Sciences Library
9101 Fayetteville Road
Raleigh, NC 27603-5696

Understanding the Human Genome Project

Second Edition

Michael A. Palladino
Monmouth University

San Francisco Boston New York
Cape Town Hong Kong London Madrid Mexico City
Montreal Munich Paris Singapore Sydney Tokyo Toronto

611.018 Pal 2006
QH 447 .P35 2006

Acquisitions Editor: Susan Winslow
Editorial Assistant: Mercedes Grandin
Marketing Manager: Jeff Hester
Production Supervisors: Shannon Tozier, Lori Newman
Production Management: Schawk, Inc.
Composition and Illustration: Schawk, Inc.
Manufacturing Buyer: Stacy Wong
Text Designer: Schawk, Inc.
Cover Designer: Seventeenth Street Studios
Text and Cover Printer: Courier/Stoughton
Cover Image: DNA and chromosomes. Jim Dowdalls/Photo Researchers, Inc.

ISBN 0-8053-4877-8

Copyright © 2006 Pearson Education, Inc., publishing as Benjamin Cummings, 1301 Sansome St., San Francisco, CA 94111. All rights reserved. Manufactured in the United States of America. This publication is protected by Copyright and permission should be obtained from the publisher prior to any prohibited reproduction, storage in a retrieval system, or transmission in any form or by any means, electronic, mechanical, photocopying, recording, or likewise. To obtain permission(s) to use material from this work, please submit a written request to Pearson Education, Inc., Permissions Department, 1900 E. Lake Ave., Glenview, IL 60025. For information regarding permissions, call (847) 486-2635.

Many of the designations used by manufacturers and sellers to distinguish their products are claimed as trademarks. Where those designations appear in this book, and the publisher was aware of a trademark claim, the designations have been printed in initial caps or all caps.

1 2 3 4 5 6 7 8 9 10—CRS—08 07 06 05
www.aw-bc.com

Contents

Introduction 1

Defining the Human Genome Project 1

Genome Basics 1

Goals of the Human Genome Project 7

How Is the Genome Studied? The Tools 9

What Have We Learned So Far? 20

Where Are All the Genes? 20

What Is This "Junk"? 22

Are Humans Really Unique? 22

Gene Discovery 24

"Take Home" Message 24

Learning About the Human Genome Project via the Internet 25

Sites to Visit and Internet Exercises 26

Genome Issues 30

Where Do We Go from Here? 34

Resources for Students and Educators 40

For Students 40

For Educators 43

Journal Articles 44

Popular Books 44

Introduction

It is indeed an exciting time to be studying biology. Over the next decade and beyond, you will witness some of the most significant biological discoveries in our history. The wealth of knowledge to be gained from the Human Genome Project will have tremendous impact in basic science and medicine in the near future. In many ways, the human genome is considered one of the great unsolved mysteries of biology—a treasure chest of genetic information. The Human Genome Project was a biological research project of unprecedented magnitude, and the challenge of identifying all the genes contained in human cells is now complete. Whether you are a biology major or nonmajor, this booklet will provide you with a basic overview of important goals and outcomes of the Human Genome Project; stimulate thought and discussion on the potential future impact of the project; and introduce some of the ethical, legal, and social implications of unraveling the keys to our genetic material. Throughout this booklet, key terms appear in bold to help you learn important concepts related to the Human Genome Project.

Defining the Human Genome Project

GENOME BASICS

A human embryo is formed shortly after fertilization—the joining of a human egg and a sperm cell. Eggs and sperm are types of cells known as sex cells or **gametes**. Contained within each gamete is nearly 6 feet of highly coiled **deoxyribonucleic acid (DNA)** packaged and condensed into a single set of 23 **chromosomes**. All the other cells in your body, such as skin cells, muscle cells, and liver cells, are known as **somatic cells** (derived from the Greek word *soma*, "body"). Over 100 trillion somatic cells are found in the human body. Each human somatic cell contains 46 chromosomes—23 chromosomes inherited from your mother and 23 chromosomes inherited from your father.

The number of chromosomes in a cell can be revealed by a **karyotype** (Figure 1). In karyotype analysis, cells (such as human white blood cells) are spread on a microscope slide, then treated with chemicals to release and stain the chromosomes. Stained chromosomes show an alternating series of light

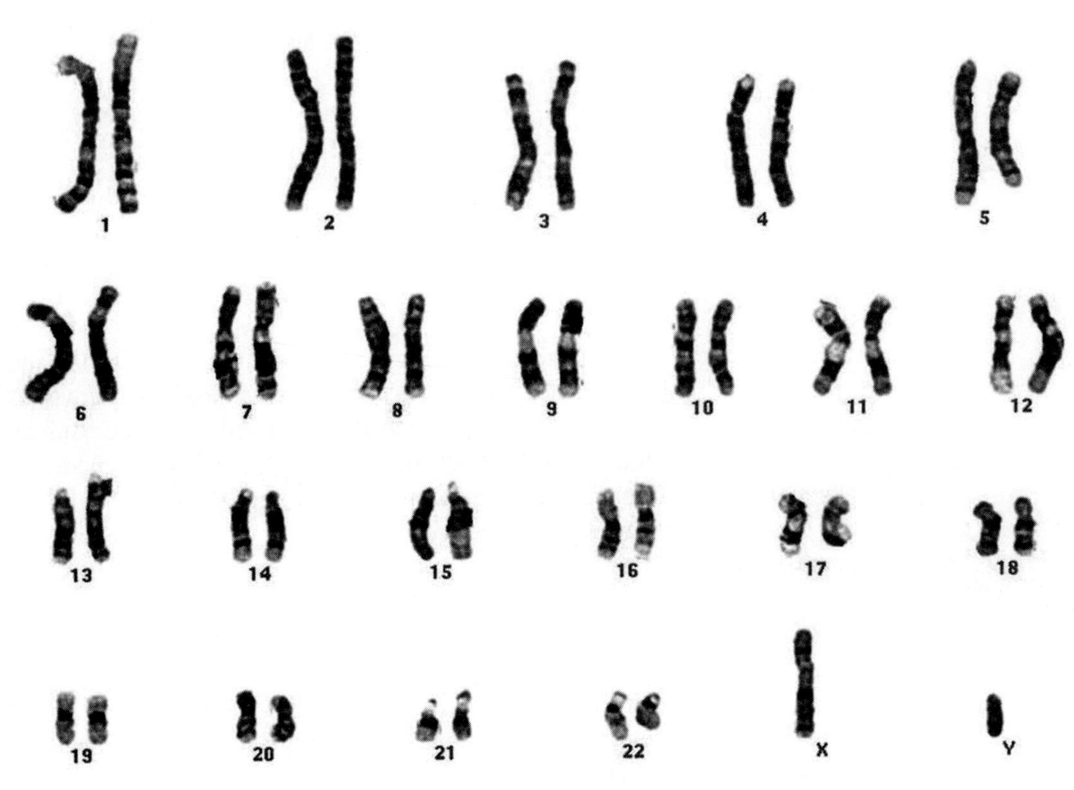

FIGURE 1 Karyotype of human chromosomes
Source: Thieman and Palladino, *Introduction to Biotechnology*, p. 33, Figure 2.7.

and dark bands. Although these bands are not visible in Figure 1, chromosomes can be aligned and paired based on their banding pattern and their size. The largest chromosome is chromosome 1. Chromosome pairs 1 through 22 are known as the **autosomes**, while the X and Y chromosomes are called the **sex chromosomes** because they are involved in sex determination. Karyotypes are frequently used to determine the sex of a fetus, and they can also be used to detect genetic abnormalities such as Down syndrome, a condition in which individuals have three copies of chromosome 21. Was the karyotype in Figure 1 produced from a somatic cell of a male or a female human? How do you know? If you have studied chromosomes before, you may already know that human female somatic cells have two X chromosomes, while somatic cells from males have one X and one Y chromosome. Notice that this karyotype is from a human male.

DNA consists of a double-stranded helix of chemical structures called **nucleotides** (Figure 2). Nucleotides are the building blocks of DNA structure.

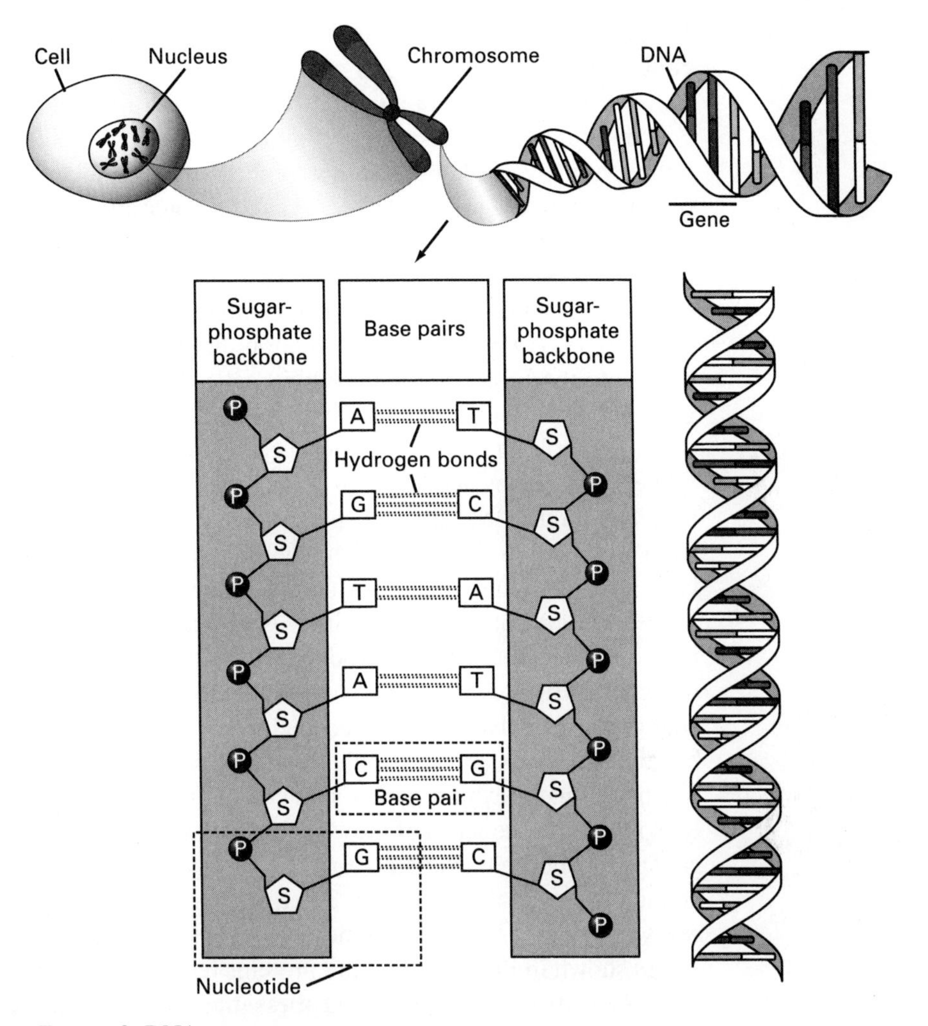

FIGURE 2 DNA structure
Source: Figures adapted from http://www.accessexcellence.org/AB/GG/dna2.html and http://www.accessexcellence.org/AB/GG/basePair2.html.

Each nucleotide consists of a phosphate group, a sugar molecule (deoxyribose sugar), and a base chemically bonded together. Four bases are found in a DNA molecule: adenine (A), guanine (G), cytosine (C), and thymine (T). The bases of DNA in two opposing strands join together by hydrogen bonding, creating **base pairs (bp)** that attach both strands of the double helix. Adenine always forms base pairs with thymine, while guanine always forms base pairs with cytosine. There are approximately 3 billion base pairs in the DNA contained within every human cell.

Contained within DNA are the instructions for life—**genes**. Most genes contain a specific sequence of nucleotides that provide cells with the instructions for the synthesis of a protein, but there are also many genes that do not encode proteins. Genes are approximately 2,000 to 4,000 nucleotides long, although many smaller and larger genes have been identified. The entire set of genes in an organism's DNA is called the **genome**. For a long time it was estimated that the human genome consisted of approximately 100,000 genes; however, as you will soon learn, this estimate has been reevaluated based on recent information from the Human Genome Project. Visit NOVA Online for "Journey into DNA," a great animation that takes you through layers of the body into the DNA contained within cells (http://www.pbs.org/wgbh/nova/genome/media/journeyintodna.swf).

By controlling the proteins produced by a cell, genes influence how cells, tissues, and organs appear, both through the microscope and with the naked eye. These inherited appearances are called **traits**. You have inherited traits from your parents by virtue of the DNA contained in your cells. Perhaps you have studied the inheritance of human traits such as eye color and skin color. Some traits are controlled by a single gene, others by multiple genes that must interact in complex ways to produce a trait. In general, traits are governed by our genes. But there is more to any trait than just genetics. Environmental conditions strongly influence gene behavior, which in turn can influence visible traits. For example, people of the same ethnic background living in warmer climates exhibit darker skin colors than people from colder climates, even though the genetics of these individuals may be very similar. Many traits go far beyond characteristics that are visible to the naked eye. Traits also include events at the molecular level, such as the metabolism of a cell—the cell's ability to manufacture molecules and utilize nutrients for energy.

As shown in Figure 3, making proteins requires that genes are copied into single-stranded molecules called **messenger RNA (mRNA)** in a complex process called **transcription**. Scientists also refer to transcription as **gene expression**. The mRNA molecules literally act as "messengers" of the genetic code by carrying information, in the form of nucleotide sequence, from the nucleus to the cytoplasm of a cell, where protein synthesis, or **translation,** occurs. **Proteins** are created by the joining together of long chains of building blocks called amino acids. The nucleotide sequence of a gene—the specific order of As, Gs, Ts, and Cs—determines which amino acids, and the amino acid composition of a protein is largely responsible for the three-dimensional shape (structure) and the function of a protein.

Because a large majority of the molecules in a cell are proteins, proteins ultimately affect the traits displayed by cells in a variety of ways. Proteins are considered the major structural and functional molecules in a cell because of the range of important roles they play. For example, some proteins function as

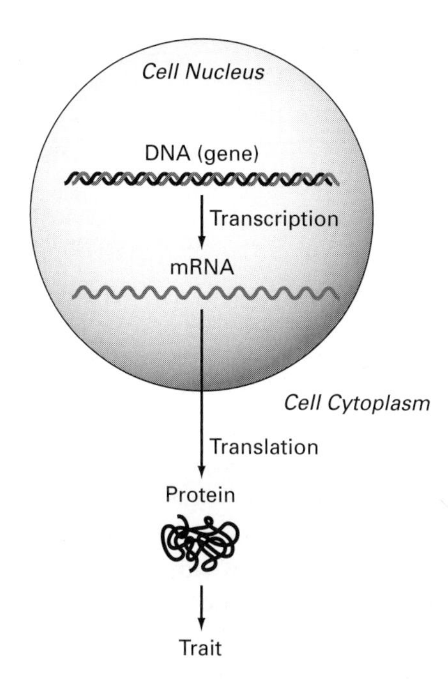

FIGURE 3 The flow of genetic information in human cells

pigments responsible for eye color, hair color, and skin color. Many proteins function as **enzymes**, proteins responsible for carrying out and accelerating chemical reactions in cells. Other proteins function as transport molecules that serve to move vitamins, minerals, hormones, and gases throughout the body. You may be familiar with an important oxygen-transporting protein in red blood cells called hemoglobin. Large classes of proteins, such as antibodies, play significant roles in our immune system and our ability to recognize and destroy foreign materials such as bacteria and viruses.

Genes can undergo genetic change. A **mutation** is a nucleotide change in the sequence of a gene. Mutations can involve large changes in genetic information or single nucleotide changes within a gene, such as changing an A to a C or a G to a T. Some mutations are due to spontaneous errors during DNA replication in cells, while others are "induced" by environmental causes, such as exposure to X-rays or ultraviolet light from the sun (that glowing tan you have is not as healthy as you might think!) and exposure to chemicals called **mutagens** that can change DNA structure. Regardless of how mutations arise, they can result in changes in protein production in cells. In some cases, mutations completely block the synthesis of a protein from a gene. In other types of mutations, a gene can still synthesize a protein, but the protein doesn't function as well as the normal protein. By affecting the proteins made by a cell, mutations can alter traits (Figure 4).

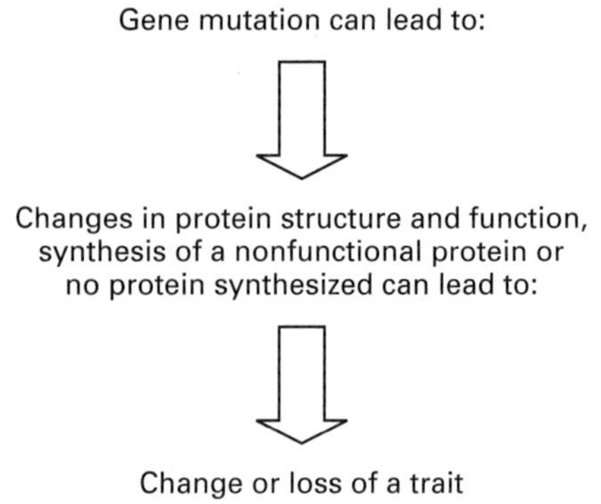

Gene mutation can lead to:

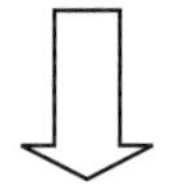

Changes in protein structure and function,
synthesis of a nonfunctional protein or
no protein synthesized can lead to:

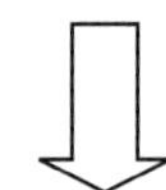

Change or loss of a trait

FIGURE 4 How mutations change protein structure and influence traits

Nucleotide changes in a gene affect proteins because the nucleotide sequence of a gene provides a cell with the information required to synthesize a protein. Change the DNA sequence of a gene, and the amino acid sequence of a protein can change. Altering the amino acid sequence of a protein often changes the overall shape of a protein, which in turn affects how the protein functions.

It might surprise you that a single nucleotide change in the genetic code of a gene can cause such a problem. A striking example of how this can happen is illustrated by the human genetic condition called **sickle-cell disease**. Sickle-cell disease results from a single mutation in the gene used to synthesize protein chains that assemble to form hemoglobin, the oxygen-binding protein contained in human red blood cells. A single nucleotide change in this gene (out of several hundred nucleotides) creates a single amino acid change in hemoglobin (out of 146 amino acids) that produces an altered hemoglobin molecule with an abnormal three-dimensional shape. While you might be thrilled to score 145 correct answers out of 146 on a biology exam, this level of accuracy in a protein can be devastating. In this case, the abnormal hemoglobin molecules bind oxygen poorly. In addition, abnormal hemoglobin changes the shape of red blood cells from disklike cells to irregularly shaped cells that take on a "sickled" look. Sickled red blood cells do not travel well through blood vessels, resulting in poor oxygenation of body tissues and other painful symptoms.

A number of conditions arise from genetic changes similar to those described for sickle-cell disease. It is likely that you have heard of cystic fibrosis, albinism, phenylketonuria, and Tay-Sachs disease. Similarly, certain behavioral and psychological illnesses result from genetic changes and can be inherited. For some genetic disease conditions to be expressed—sickle-cell disease included—a person must have inherited two chromosomes carrying the same defective gene. If only one chromosome contains a mutation for a given gene, a person is likely to be normal for that trait, because one chromosome with a "good," nonmutated copy of the gene can often produce enough protein to override the effects of the defective gene. Individuals with one defective and one normal copy of a gene are commonly called **carriers** because they can pass ("carry") a defective gene to offspring but will typically never show the diseased trait themselves. Remember this concept for our later discussion concerning genetic testing and genetic privacy. Now that we have considered some of the basics of human genetics, let's explore the purpose of the Human Genome Project.

GOALS OF THE HUMAN GENOME PROJECT

In 1990, the Office of Health and Environmental Research of the U.S. Department of Energy (DOE) proposed the Human Genome Project as a joint venture between the National Institutes of Health (NIH) and DOE, with a 15-year time frame for completion. You might wonder why the DOE was involved. DOE had long been interested in studying genetics, in part because of their research on the effects of nuclear radiation on human genetics. The project funded seven major sequencing centers in the United States. Over time the project grew to become an international effort with contributions from scientists in 18 countries, but the work was primarily carried out by the International Human Genome Sequence Consortium, which involved nearly 3,000 scientists working at 20 centers in six countries: China, France, Germany, Great Britain, Japan, and the United States. The estimated budget for completing the genome was $3 billion, a cost of $1 per nucleotide. Driven in part by competition from private companies, the Human Genome Project turned out to be a rare government project that completed all of its initial goals, and several addition goals, more than 2 years ahead of schedule and under budget!

One of the most aggressive competitors on the project was a private company known as Celera Genomics—aptly named from a word meaning "swiftness." The company was swift indeed when in 1998 it announced its intention to use novel technologies to sequence the entire human genome in three years! Fearful of how private corporations might control the release of genome information, U.S. government groups involved in the Human Genome Project were effectively forced to keep pace with private groups to stay competitive in the race to complete the genome. Interestingly, in July 2005, Celera Genomics released its formerly proprietary human, mouse, and rat genome sequences to the public domain.

In 1998, as a result of accelerated progress, a revised target date of 2003 was set for completion of the project. On June 26, 2000, leaders of the Human Genome Project and Celera Genomics participated in a press conference with President Clinton to announce that a rough "working draft" of approximately 95% of the human genome had been assembled (nearly 4 years ahead of the initially projected timetable). At a joint press conference on February 12, 2001, Dr. Francis Collins, director of the NIH National Human Genome Research Institute and director of the Human Genome Project, and Dr. J. Craig Venter, director of Celera Genomics, announced that a series of papers describing the initial analysis of the genome working draft sequence were to be published by their research groups in the prestigious journals *Nature* and *Science,* respectively.

Scientists spent the next two years working to fill in thousands of gaps in the genome by completing the sequencing of pieces not yet finished, correcting misaligned pieces, and comparing sequences to ensure the accuracy of the genome. Finally, on April 14, 2003, the International Human Genome Sequencing Consortium announced that its work was done. A "map" of the human genome was essentially complete with virtually all bases identified and placed in their proper order with the exception of about 300 relatively small gaps of DNA that remain problematic but will be completed shortly.

What were the goals and objectives of the Human Genome Project? Specifically, the project was designed to do the following:

- Create genetic and physical maps of the 24 human chromosomes (22 autosomes, X and Y chromosomes).
- Identify the entire set of genes in the DNA of human cells. This included mapping each gene to its chromosome and determining the sequence of each gene. Originally thought to be around 100,000 genes when the project started, the total number of human genes is much less than this, as you'll soon learn.
- Determine the nucleotide sequence of the estimated 3 billion base pairs of DNA that comprise the human genome.
- Analyze genetic variations among humans. This included the identification of single-nucleotide polymorphisms (SNPs).
- Map and sequence the genomes of **model organisms,** including bacteria, yeast, roundworms, fruit flies, and mice.
- Develop new laboratory and computing technologies (including databases of genome information) that can be used to advance our analysis and understanding of gene structure and function.
- Disseminate genome information among scientists and the general public.
- Consider ethical, legal, and social issues that accompany the Human Genome Project and genetic research.

Although many scientists were skeptical that the project would succeed and were critical of the large amounts of money allocated for the project, the

Human Genome Project has been described by others as a research effort guaranteed to succeed, because so many scientists believed it was eventually possible to determine the sequence for all 3 billion base pairs in the human genome. What has surprised nearly everyone has been the speed with which many of the project's goals were achieved.

With completion of the Human Genome Project, scientists now have access to the "book of life." Yet, as you will learn in this booklet, completion of the Human Genome Project is just the tip of the iceberg in our understanding of genes and human genetics. How can information from the genome project be used? Gaining a better understanding of normal gene structure and function is one great benefit of the project. However, one of the most exciting outcomes will be a greater ability to identify and treat genetic disorders. For instance, new knowledge from the Human Genome Project will lead to the development of advanced screening techniques for diseased genes, early warning, and even prevention and treatment of human genetic diseases through development of genetics-based therapies, including gene therapy—topics we will discuss shortly.

How Is the Genome Studied? The Tools

Studying the genome of any organism is possible because of a range of sophisticated and elegant laboratory techniques in molecular biology. Most genome studies begin with isolating DNA from a tissue sample of interest. In the Human Genome Project, DNA was primarily isolated from human blood samples. Adult human red blood cells cannot be used as a source of DNA because they lack a nucleus and therefore do not contain DNA. Instead, white blood cells—important cells in the immune system that do contain a nucleus—can be removed from a blood sample and used to isolate DNA.

Isolating chromosomal DNA from blood cells is relatively simple. You may have the opportunity to isolate DNA in one of your biology courses. Once DNA is obtained, however, it is not possible to sequence an entire chromosome at once. Whole chromosomes are simply too large. Most chromosomes average around 150 million base pairs in size! The largest human chromosome, chromosome 1, is ~260 million base pairs. Even the smallest human chromosome, the Y chromosome, is ~60 million base pairs in size. A chromosome must be broken into small pieces that can be easily manipulated to identify genes and ultimately to determine the nucleotide sequence of the entire chromosome.

Cutting a chromosome down to size is accomplished using DNA-cutting enzymes called **restriction enzymes**. Restriction enzymes are essentially "scissors" for molecular biologists. A quick look in the freezer of any molecular biology lab will reveal dozens of restriction enzymes. These enzymes are isolated from bacteria and given abbreviated names based on the bacteria they come from. For example, one of the first enzymes identified and most widely used is called *Eco*RI. *Eco*RI is isolated from *Escherichia coli*, the bacterium naturally found in the intestines of animals, including humans. Restriction enzymes have the ability to cut DNA strands at specific sequences of base pairs (Figure 5).

Restriction enzymes allow scientists to cut DNA into small pieces that can be copied—a technique called **gene cloning**. A wide range of gene cloning techniques exist.

In gene cloning, the DNA pieces of interest can be spliced into other pieces of DNA, called **vectors,** that are used to carry and replicate the DNA piece of interest. Scientists cut and splice human DNA pieces into a variety of different vectors depending on the size of the DNA piece to be cloned. This must be done because human chromosomes are generally too large to manipulate and clone without breaking them up into smaller fragments of a more manageable size. Yeast artificial chromosomes (YACs), bacterial artificial chromosomes (BACs), cosmids, and plasmids (small circular pieces of DNA found in bacteria) all serve as vectors that can be used to clone pieces of human DNA. Each vector is limited by the size of DNA fragments that can be inserted into the vector. Details of these vectors are not important, but you are likely to encounter these terms as you review some of the Web sites described in this booklet.

Once a human DNA piece has been inserted into a vector, the vector can be used to make more copies (or clones) of the human DNA piece. To accomplish this, vectors are usually placed into bacteria or yeast for DNA cloning. These microorganisms are used because they are easy and cheap to grow in the lab, and they are pretty good at replicating DNA, even if the DNA is from a human source. An illustration of cloning a piece of a human chromosome is shown in Figure 6. As shown in this figure, one strategy for cloning human chromosomes involves using restriction enzymes to cut DNA fragments into fragments of varying sizes, from larger to smaller, which are subsequently cloned into small vectors (plasmids) that allow the DNA to be sequenced.

DNA sequencing involves determining the exact arrangement of DNA nucleotides—the specific order of As, Gs, Ts, and Cs—in a piece of DNA. A vari-

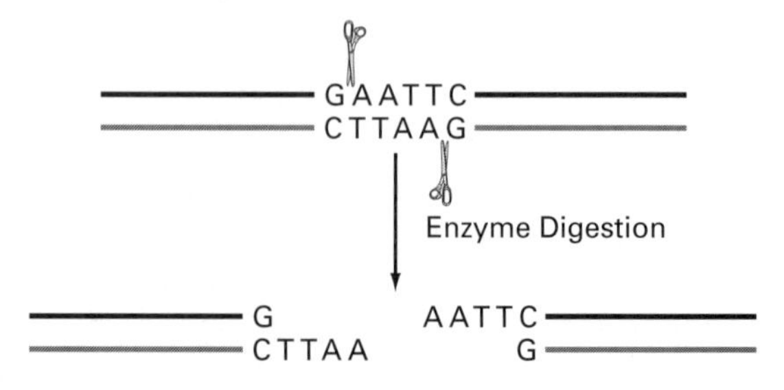

FIGURE 5 The action of restriction enzymes

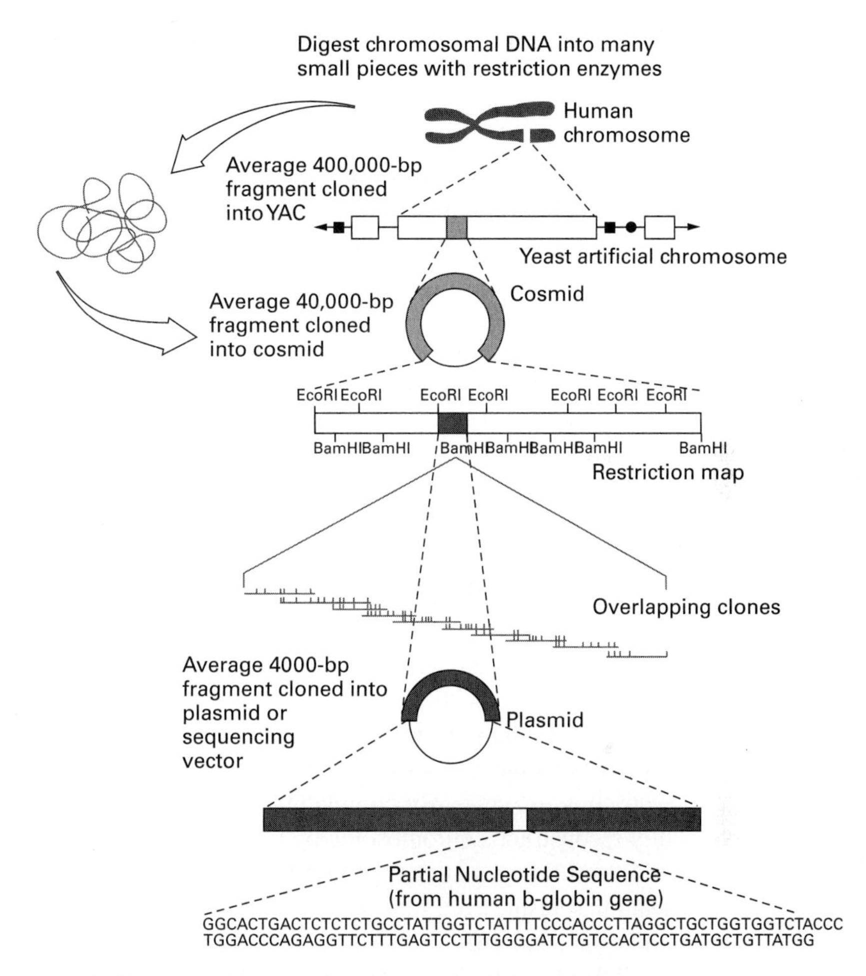

FIGURE 6 Cloning and sequencing pieces of a human chromosome
Source: Thieman and Palladino, *Introduction to Biotechnology,* p. 265, Figure 11.21.

ety of sequencing techniques exist, but most use DNA nucleotides that have been tagged with different-colored dyes that fluoresce when exposed to a laser. Reactions are carried out to copy a DNA sequence in the presence of these modified nucleotides, and then the sequence can be read based on the fluorescence pattern resulting from illumination of DNA samples with a laser beam. Progress of the Human Genome Project was greatly accelerated by the development of fast-paced computer-automated sequencing machines that work around the clock to generate large amounts of sequence data. As sequencing technology has continued to improve, there have been some scientists who have suggested that in the future it may be possible for a person to have his or her genome sequenced for as little as $1,000! Once DNA has been sequenced, computer programs are used to catalog the sequence information; interpret the sequence to determine if it contains protein-coding instructions; and compare it to databases of known sequences to find out if this sequence has already been determined, if it represents a novel piece of chromosome sequence, or if it shares similar sequences with other DNA pieces. The use of computer hardware and software for sequence analysis, along with many other applications such as archiving sequences in databases to store, share, and compare DNA and protein data, is part of a relatively new discipline called **bioinformatics,** an integrated field involving biology and information technology.

The approach we have just discussed is a very basic overview of one of the primary methods used to clone and sequence human DNA. But if a single chromosome contains millions of base pairs, how can all this sequence data be assembled to construct a sequence of an *entire* chromosome? A challenging task indeed!

One way to sequence an entire chromosome involves a random cloning process called "shotgun" cloning and sequencing. In this process, chromosomal DNA is cut into smaller pieces with several different restriction enzymes. For example, *Eco*RI and another enzyme called *Bam*HI can be used. The different pieces generated are cloned and sequenced as we just discussed. The idea behind this approach is to randomly generate and sequence short pieces of DNA fragments with the hope of sequencing short overlapping pieces. High-powered computer programs are then put to work to look for and align overlapping sections of DNA sequences from individual pieces cut with the two restriction enzymes. By completing this genetic puzzle, it is possible to literally reconstruct a chromosome by walking along these overlapping fragments and assembling a stretch of continuously overlapping sequences (Figure 7). Data that result from this type of study produce a map of overlapping pieces and restriction-enzyme cutting sites on a chromosome called a **physical map**.

Before cloning and sequencing technologies were available to identify and map genes, scientists often studied large numbers of families with a history of a particular genetic disorder and created genetic maps based on intricate patterns of inheritance and other genetic data. Genetic maps are still used when studying disease genes to provide supporting data for physical mapping studies. Physical maps provide the molecular details of a gene and its location.

Visit the NOVA Online "Sequence for Yourself" site at http://www.pbs.org/wgbh/nova/genome/media/sequence.swf for informative animations on DNA cloning and sequencing and assembling cloned DNA fragments to create a physical map of a chromosome.

In 1995, a shotgun-cloning strategy was used by scientists at The Institute for Genomic Research to sequence the 1.8 million base pairs in the genome of a strain of bacteria called *Haemophilus influenza.* This was the first time an entire genome for any organism had been sequenced. Many doubted that shotgun-cloning strategies could be effectively used to sequence larger

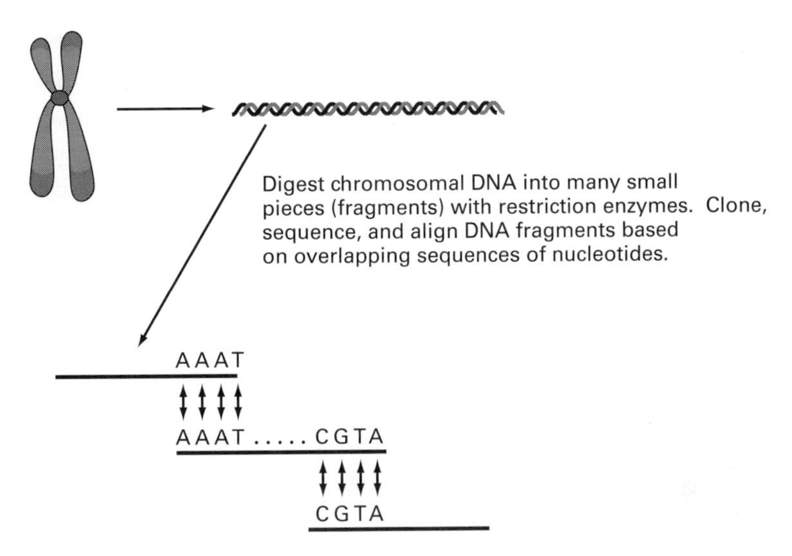

FIGURE 7 Aligning pieces of human chromosomal DNA to create a chromosome map

genomes, but development of this technique and novel sequencing strategies rapidly accelerated progress of the Human Genome Project.

While shotgun approaches are very effective for sequencing and mapping segments of a chromosome, such approaches are not practical for identifying expressed genes because, as you will learn shortly, only a very small percentage of human DNA consists of genes used to make proteins. Much of our genome contains non-protein-coding DNA. In some ways, using a shotgun approach to identify gene sequences in the genome is like searching for a needle in a haystack. A better approach for identifying genes involves techniques using mRNA. Recall that mRNA is a copy of a gene that is used to make a protein. By working with mRNA, scientists are studying expressed genes in a tissue and not looking at non-protein-coding DNA. The mRNA can be copied into DNA. This DNA is now called **complementary DNA,** or **cDNA,** because it is identical to a sequence of mRNA. Pieces of cDNA can be sequenced to produce fragments called **expressed-sequence tags,** or **ESTs**. ESTs represent small pieces of DNA sequenced from genes that are expressed in a cell. Rarely do ESTs span an entire gene, but these small pieces can be used as "tags" to ultimately determine the sequence of an entire gene. ESTs have played an important role in the identification of human genes.

Figure 8 shows a "big picture" representation of how DNA sequence can be determined from restriction fragments of DNA and how overlapping restriction fragments can be assembled to create a physical map of a chromosome. Notice that partial maps of chromosome 19 are shown, including the location of human disease genes for a type of diabetes (insulin-resistant diabetes) and an

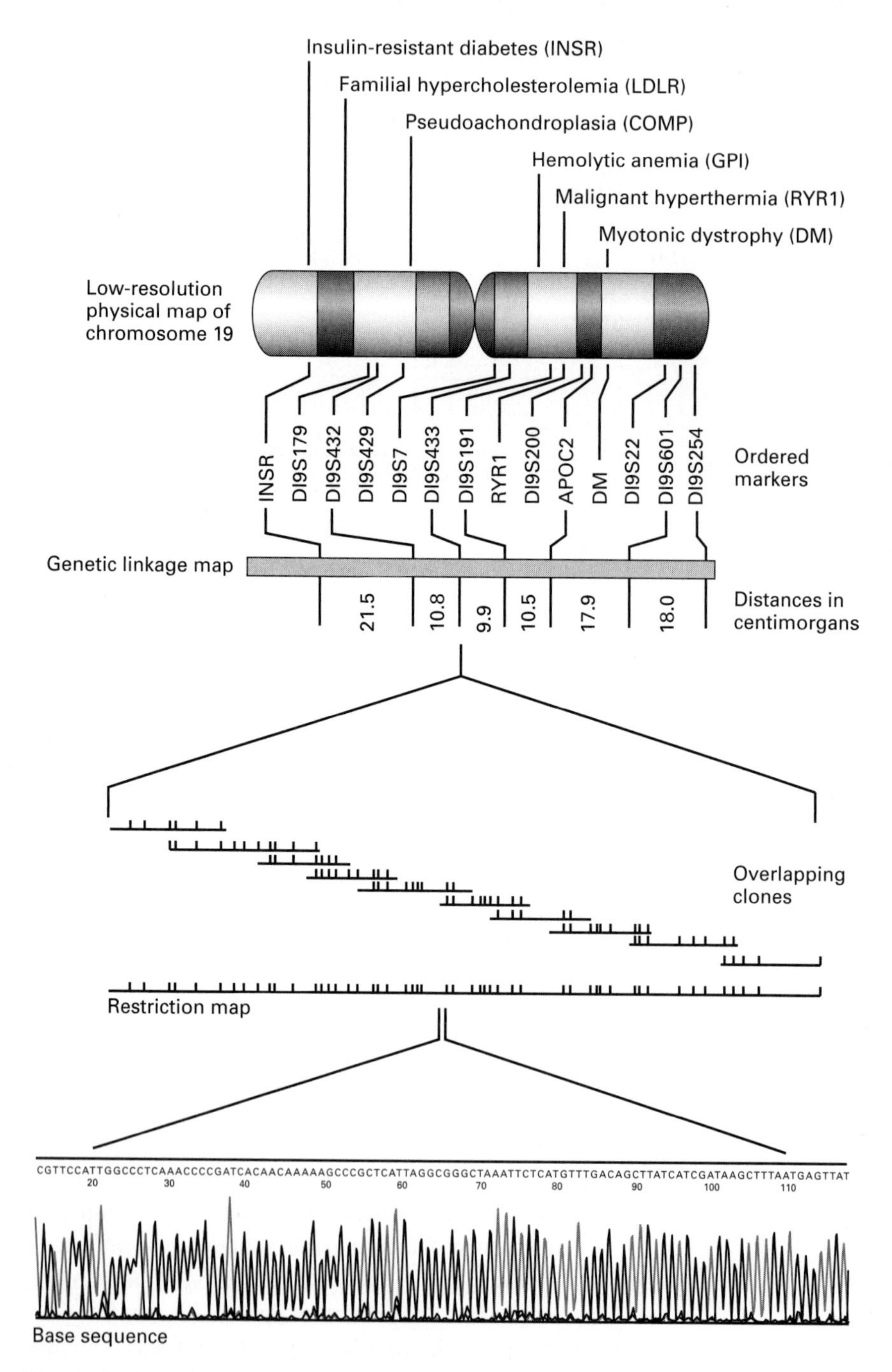

FIGURE 8 Partial gene maps and DNA sequence for human chromosome 19
Source: Printed with the permission of the U.S. Department of Energy Human Genome Program. Adapted from Figure 3, p. 11 from "To Know Ourselves." Human Genome Program, U.S. Department of Energy, To Know Ourselves, 1996.

inherited condition in fat metabolism that results in high blood cholesterol and increased risk of heart disease (familial hypercholesterolemia).

As you will learn when we discuss how the Internet can be used to help us find out more about the genome, scientists have assembled maps of human chromosomes with the locations of all known normal and disease genes.

Manipulating DNA to map genes involves some very powerful and sophisticated science. I expect that you may have more than a few questions about how the genome is studied. Shortly, you'll be introduced to Web sites that may provide answers to many of your questions. Meanwhile, here are a few questions to consider.

Whose genome is being sequenced, and how do we know it is representative of all humans?

Obviously, it is not possible to collect and sequence DNA from every living human. So how can we say the human genome has been determined? In some genome centers, DNA was collected from many different individuals. Scientists at Celera Genomics worked with DNA from five individuals: two males and three females self-identified as Caucasian, African American, Hispanic, or Asian.

What is a truly normal genome? With the exception of identical twins, the genome of each person is unique. All humans share a basic set of genes that control human development and normal body functions. It is not possible to determine a sequence that is an exact match for any one person's genome. Because there are thousands of variations in the range of different human traits, there is no one "normal" genome sequence.

Rather, you should think of the Human Genome Project as an effort to produce an initial sequence as a typical "reference" sequence of our genome. This reference sequence will be a resource for evaluation and comparison to help scientists understand human genetic similarities and, perhaps more importantly, the range of genetic variation between individuals. The reference sequence will tell us what genes are in the human genome and what some of the most common variations are, but it will not be possible to identify all variations of all genes in humans without sequencing the genome of every human.

By sequencing the genome from individuals of different races, scientists are assembling a picture of what a representative (reference) genome is, and over time this reference genome will be expanded as we learn more about genetic variations among humans. But without question, we now know that virtually all of DNA in humans of all backgrounds is exactly the same. The Human Genome Project has revealed that 99.9% of the nucleotide sequence of the human genome is exactly the same in all humans.

If all cells have the same DNA, then what makes a skin cell different from a muscle cell or a neuron different from a liver cell?

After fertilization, each individual begins life as a single cell. During development of the embryo, how do some cells know they should become skin, while other cells form the tissues of the kidneys or muscle or brain? These are questions with complex answers. Part of the answer lies in understanding the genes that control cell and tissue development and how they interact with one another. There are a number of genes that are tissue specific, meaning they are expressed by certain tissues but not others. Tissue-specific genes are part of the genetic "programming" that tells some cells to develop into skin and some cells to become muscle. One of the real challenges following the Human Genome Project will be to learn how tissue-specific genes function and interact with each other to trigger organ development. A better understanding of the cellular events and genes involved in tissue development may lead to a greater ability to produce tissues and organs for wound healing and transplantation in humans.

A major goal of the Human Genome Project was to sequence the genomes of "model" organisms, such as flies, yeast, and worms. Why do scientists care about the genomes of these other organisms?

Model organisms are critically important to scientists because we can't do many of our sophisticated genetic studies on humans. It is unethical and illegal to force humans to breed or to insert or remove genes into humans to learn how human genes function! However, these approaches are widely used to study genes in model organisms. Mice, yeast, fruit flies, worms, and even the zebrafish (a common fish in home aquariums) have all played important roles in our understanding of human genetics. *Arabidopsis thaliana*, a small, flowering plant related to cabbage, is also widely used as a model organism in plant biology. Although not of major agricultural value, *Arabidopsis* offers many advantages for basic research in genetics and molecular biology.

Important genes are highly conserved from species to species. This is one reason why model organisms are so valuable. If we can identify and learn about important genes in model organisms, we can form hypotheses and make predictions about how these genes may function in humans. Comparing segments of DNA between different species as a way to compare similarities and differences in DNA and to identify genes and gene regulatory sequences is a field of genetics known as **comparative genomics**. See Table 1 to compare the genome size of humans and several model organisms.

Many genes in different model species have been shown to be related to human genes based on DNA sequence similarity (see Figure 9). Genes with sequence similarity are called **homologs**. For instance, the **obese *(ob)* gene,** which produces a protein hormone called leptin, was first discovered in mice. Eating stimulates fat cells to produce leptin, which affects cells in the brain to suppress appetite and stimulate metabolism of food. Other studies have shown that mice deficient in the leptin gene grow dramatically overweight (see Figure 10).

TABLE 1. Comparison of Selected Genomes

Organism (scientific name)	Approximate size of genome (date completed)	Number of genes	Approximate percentage of genes shared with humans	Web access to genome databases
Bacterium (*Escherichia coli*)	4.1 million bp (1997)	4,403	Not determined	www.genome.wisc.edu/
Chicken (*Gallus gallus*)	1 billion bp (2004)	~20,000–23,000	60%	http://genomeold.wustl.edu/projects/chicken
Dog (*Canis familiaris*)	6.2 million bp (2003)	~18,400	75%	http://www.ncbi.gov/genome/guide/dog
Chimpanzee (*Pan troglodytes*)	~3 billion bp (initial draft, 2005)	~20,000–24,000	96%	http://www.nature.com/nature/focus/chimpgenome/index.html
Fruit fly (*Drosophila melanogaster*)	165 million bp (2000)	~13,600	50%	www.fruitfly.org
Humans (*Homo sapiens*)	~2.9 billion bp (2004)	~20,000–25,000	100%	www.doegenomes.org
Mouse (*Mus musculus*)	~2.5 billion bp (2002)	~30,000	~80%	www.informatics.jax.org
Plant (*Arabidopsis thaliana*)	119 million bp (2000)	~26,000	Not determined	www.arabidopsis.org
Rat (*Rattus norvegicus*)	~2.75 billion bp (2004)	~22,000	80%	www.hgsc.bcm.tmc.edu/projects/rat
Roundworm (*Caenorhabditis elegans*)	97 million bp (1998)	19,099	40%	genomeold.wustl.edu/projects/celegans
Yeast (*Saccharomyces cerevisiae*)	12 million bp (1996)	~5,700	30%	genomeold.wustl.edu/projects/yeast.index.php

Source: Nature Genome Gateway Web site www.nature.com/genomics/papers/.

```
Human ob gene
gtcaccaggatcaatgacatttcacacacg---tcagtctcctccaaacagaaagtcacc
||||||||||||||||||||||||||||||   || || ||| |||| ||||  ||||| 
gtcaccaggatcaatgacatttcacacacgcagtcggtatccgccaagcagagggtcact
Mouse ob gene

ggtttggacttcattcctgggctccaccccatcctgaccttatccaagatggaccagaca
|| |||||||||||||||||||| |||||||| ||||  || ||||||||||||||||| 
ggcttggacttcattcctgggcttcaccccattctgagtttgtccaagatggaccagact


ctggcagtctaccaacagatcctcaccagtatgccttccagaaacgtgatccaaatatcc
||||||||||| |||||| ||||||||||  ||||||||  ||| ||| | || ||| ||
ctggcagtctatcaacaggtcctcaccagcctgccttcccaaaatgtgctgcagatagcc
```

FIGURE 9 Comparison of the human and mouse *ob* genes
Partial sequences for these homologs are shown with the human gene on top and the mouse gene sequence below it. Notice how the nucleotide sequence for these two genes is very similar as indicated by the vertical lines between identical nucleotides. Source: Sequence comparison conducted by BLAST analysis at: http://www.ncbi.nlm.nih.gov/BLAST/.

Although the genetics of weight control involves much more than just one gene, subsequent discovery of a human homolog for leptin has led to a new area of research with great promise for providing insight on fat metabolism in humans and the genetics that may influence weight disorders.

Historically, significant scientific discoveries in almost all fields of biology, including anatomy and physiology, biochemistry, cell biology, developmental biology, genetics, and molecular biology, were first made in model organisms and then related to humans. For instance, in most developing embryos, some cells have to die to make room for others. How does the body know where to develop certain organs and determine which cells must die to make room for others?

Answering these important developmental questions has been greatly advanced by the roundworm *Caenorhabditis elegans*. The adult *C. elegans* has 959 cells. Maps of *C. elegans* have been created that allow scientists to determine the fate (lineage) of all 959 cells to form the nervous system, intestine, and other tissues of the worm. Of these cells, 131 are destined to die in a form of cell suicide known as programmed cell death, or *apoptosis*. During development of a human embryo, sheets of skin cells create webs between the fingers and toes; apoptosis is responsible for the degeneration of these webs prior to birth. But apoptosis is significant in other ways. We now know, for example, that apoptosis is involved in neurodegenerative diseases such as Alzheimer's disease, Huntington's disease, amyotrophic lateral sclerosis (Lou Gehrig's disease), and Parkinson's disease, as well as other conditions such as arthritis and forms of infertility. How might we better understand the genes involved and slow or stop these degenerative processes? Model organisms will help us answer these questions.

How do cells know if they are to become heart or liver, and how do they know how to move and where to move to settle and form an organ? How does an embryo know where and how to form limbs? Model organisms were also used to identify another important set of human developmental genes called the homeobox genes. This cluster of genes provides instructions that tell the developing embryo where to position cells to form limbs. Mutations in human homeobox genes can result in infants being born with missing fingers or extra fingers.

Recently researchers at the University of Utah have initiated a project to determine the genome of planaria *(Schmidtea mediterranea)*, small translucent aquatic flatworms that are commonly studied in high school and college biology labs. One reason why the planaria genome is of interest is that cells from these organisms show little signs of cell aging, cells that do age are continually replaced, and planaria have incredible abilities to regnerate tissues when damaged.

It is possible to learn a great deal about gene function in model organisms in ways that are not possible to study in humans. For instance, gene **knockout mice** can be created in which a particular gene is disrupted to make it nonfunctional, and the effect of losing the function of that gene is studied. For example, as shown in Figure 10, knocking out the *ob* gene in mice provided researchers with compelling evidence that the *ob* gene is clearly involved in obesity.

As another example, disrupting a gene involved in fat metabolism in the *C. elegans* has been shown to dramatically alter the life span of these worms,

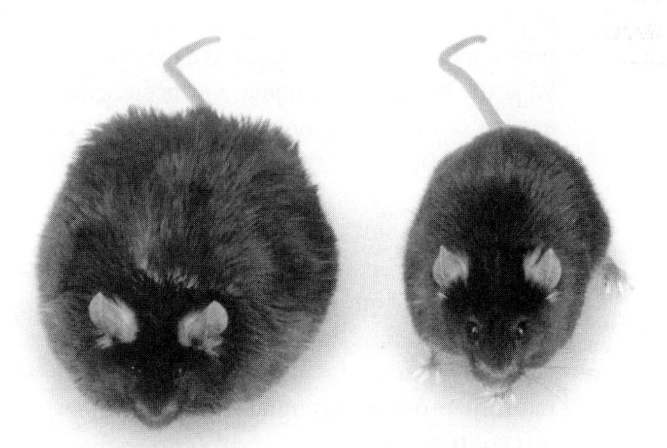

FIGURE 10 Knocking out the *ob* gene creates obese mice
The ob knockout is on the left and a normal litter mate is on the right.
Source: John Sholtis-Amgen, Inc.

and we now know that a similar gene is found in humans. Extra copies of a gene can also be inserted in mice to study the effects of overexpression of a gene. Gene addition has been used in mice and flies to reverse the effects of damaged genes that cause altered traits. As you will learn later in this booklet, model organisms are being used to develop gene therapy strategies that will be applied to humans to treat genetic diseases.

As you can see, model organisms offer much to help us learn about human genetics. In the next section, we will continue to see the importance of model organisms when we consider what genes have been identified in the genomes of both humans and model organisms.

What Have We Learned So Far?

In 2004, the completed sequence published by the International Human Genome Sequencing Consortium reported a genome of 2.85 billion nucleotides. Eventually when all of the small gaps are fixed it is likely that the "final" size of the genome will be around 3.1 billion nucleotides. The genome was sequenced with an accuracy of ~99.999%. This equates to one mistake every 100,000 base pairs. About 10 times more accurate than initially expected!

A great deal about the human genome has been revealed in a relatively short period of time, although many of the greatest challenges and discoveries lie ahead. This section briefly highlights some of the major discoveries of the Human Genome Project.

WHERE ARE ALL THE GENES?

On February 12, 2001, the International Human Genome Sequencing Consortium and Celera Genomics held a joint press conference to announce that nearly 95% of the human genome had been sequenced. One of the most surprising aspects of this press conference was a new estimation that the genome consists of only 30,000 to 40,000 genes and not 100,000 genes, as previously predicted.

By 2004 this number had been revised to approximately 20,000 to 25,000 protein-coding genes. The prediction of 100,000 genes was based primarily on estimates that human cells make approximately 100,000 to 150,000 proteins. One reason why the actual number of genes is so much lower than the predicted number is the discovery of large numbers of gene families with related functions. In addition, it has been found that many genes can code for multiple proteins through a complex process of mRNA processing known as **alternative splicing**.

Analysis of human genes by functional categories has provided genome scientists with a snapshot of the numbers of genes involved in different molecular functions. Figure 11 shows one proposed interpretation of assigned functions to human genes based on gene sequence similarity to genes of known function. Not surprisingly, many genes encode enyzmes, while other large categories of genes encode proteins involved in signaling and communication within and between cells and DNA and RNA binding proteins. Notice that this estimate also shows that approximately 42% of human genes have no known function, although recent evidence suggests that functions for over half of our genes remains unknown! Keep this in mind if you are interested in a career in genetics research, because understanding what these genes do will provide exciting career opportunities for many years into the future.

Most human genes average around 3,000 base pairs in size, and the largest gene, called the dystrophin gene, is ~2.4 million base pairs in size. Mutations in this gene are responsible for the muscle disorder muscular dystrophy. Not surprisingly, the largest chromosome, chromosome 1, has the most genes, approximately 3,000 genes. The smallest chromosome, the Y, has the fewest genes, ~250 genes.

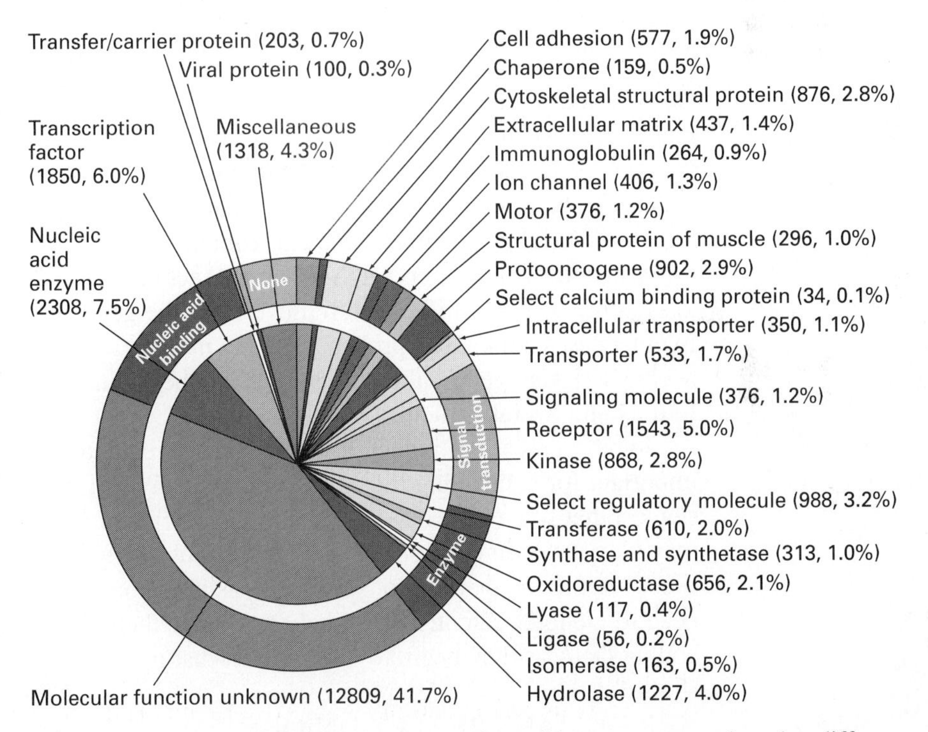

FIGURE 11 Proposed functions for the numbers of human genes assigned to different functional categories. Parentheses show percentages based on the 2001 published data by Venter et al. for 26,383 genes.
Source: From Venter, J. C., et al. (2001). The sequence of the human genome. *Science*, 291, 1304–1351, Figure 15.

WHAT IS THIS "JUNK"?

As a result of the Human Genome Project, it is now known that over 95% of our DNA consists of non-protein-coding DNA—DNA that does not contain genes. Because this DNA does not contribute directly to our traits, it has often been referred to as "junk DNA." Does it surprise you to learn that most of the DNA in your cells is junk? Actually, this DNA is not junk at all. This tongue-in-cheek expression was coined because of our lack of understanding of what non-protein-coding DNA does, not because this DNA is unimportant.

Over 50% of the non-protein-coding DNA consists of a variety of repeat sequences—sequences of repeating base pairs (such as GCGCGC) that have been duplicated in large numbers and randomly scattered throughout the genome. Many of these sequences are ancient sequences of DNA that have existed in the human genome for well over 400 million years. Repeat sequences truly represent fossil records of our genetic past. By learning about other species that contain similar sequences, we can begin to better understand where these sequences came from and how they became part of the human genome through evolution.

For example, a short sequence (approximately 300 base pairs) of junk DNA called *Alu* appears almost a million times in our genome. Recent estimates suggest that *Alu* sequences may comprise nearly 7% of our entire genome. Why do we have these sequences? In some cases, junk DNA appears to be remnants of older genes that are no longer used, and perhaps there is no good reason to eliminate it from the genome. Some scientists have suggested that certain junk DNA sequences act in a "selfish" manner because in fact they may be difficult to remove from the genome through evolution. In other cases, junk DNA probably plays a number of structural roles by connecting adjacent genes, allowing for spacing between genes, and creating the folding and overall structure of chromosomes.

Some segments of junk DNA also contain regulatory sequences that are important for controlling the expression of many genes. Recent studies also suggest that some junk DNA is used to make regulatory RNA molecules that may play a role in controlling gene expression. So even if geneticists sometimes refer to the majority of your DNA as junk, that doesn't mean you don't need it! Genome scientists are likely to change their perspective on this non-coding DNA as they begin to understand what it really does.

ARE HUMANS REALLY UNIQUE?

We think of ourselves as unmatched by other species for our ability to communicate through speech and writing as well as many other attributes, including walking upright, creating music, making a good pizza, and exploring distant planets. Humans take pride as a unique, superior species. But are we really so unique? At the genetic level, the answer to this question may surprise you. Genetically, we are not so different from many other species—species often considered inferior to humans.

From yeast and bacteria to mice, we share large numbers of genes with these species. As we have discussed, one of the goals of the Human Genome Project was to develop a better understanding of genetic similarities and differences between humans and other species, particularly other mammals. This understanding has been fostered in part by studying model organisms.

The Human Genome Project has shown us that we share a large number of genes with other organisms—proof that there is great unity of life even between seemingly very different organisms and providing solid evidence for the evolution of genes from species to species. Can you believe that you share approximately 50% of your genes with the pesky fruit flies that you bring home on fruit from the grocery store? We share nearly 40% of our genes with roundworms and 30% of our genes with yeast—the same yeast we use to help make bread rise and ferment alcoholic beverages. We share even more genes with mice and chimpanzees; approximately 80% and 96% of our genes, respectively, are similar in structure and function.

Recently, the genome for "man's best friend" was completed, which revealed that we share about 75% of our genes with dogs. Human DNA even contains around 100 genes that are also present in many bacteria. However, humans do make many more proteins than most model organisms. Refer to Table 1 for a comparison of the genomes from model organisms and humans. The Web sites presented in the table are excellent resources for learning about the genomes of model organisms.

Our genetic relatedness to model organisms is evidence of our evolutionary past. What can we learn from our inappropriately called "lower" relatives? If genes found in bacteria, fruit flies, yeast, mice, and other organisms are also found in humans, doesn't this suggest that these genes must be pretty important? Many genes that determine our body plan, organ development, and eventually our aging and death are virtually identical to genes in fruit flies. Mutated genes that are known to give rise to disease in humans also cause disease in fruit flies. According to a report from the Howard Hughes Medical Institute *(The Genes We Share with Yeast, Flies, Worms, and Mice: New Clues to Human Health and Disease)*, approximately 61% of genes mutated in 289 human disease conditions are found in the fruit fly. This group includes the genes involved in prostate cancer, pancreatic cancer, cardiac disease, cystic fibrosis, leukemia, and a host of other human genetic disorders.

It may seem hard to believe that it takes only about twice as many genes to make a human as it does a fruit fly (~25,000 versus 13,600). And plants, such as rice, have even more genes than we do! Who would argue that a rice plant is more complex than a human? But in fact plants possess many complex aspects of metabolism that humans lack such as using photosynthesis to convert energy from the sun into chemical energy. What and who we are is clearly much more than the total number of genes we have. Being human is far more complicated that simply possessing a human genome—we are infinitely more complex than just the sum of our genes and body parts. What defines us genetically is the complexity of how our genes are used and how these genes interact with one another to provide the myriad functions and unique characteristics of the creatures we call humans. This is what distinguishes us from other species.

GENE DISCOVERY

As expected, the Human Genome Project has led to the discovery of new genes involved in a wide range of activities in human cells. The total number of human genes is smaller than expected in part because of duplicate genes and gene families—genes with related functions. By grouping genes into families, it becomes easier to determine how related genes work and identify important functions they may have. Not surprisingly, many of the newly discovered genes are involved in complex reactions that control body metabolism. Another large group of genes control the expression of other genes. Still others are responsible for maintaining the shape, structure, and function of body tissues.

As genome scientists evaluate the functions of newly discovered genes, they begin to identify genes that not only control normal functions but also may be connected to human disease conditions. These discoveries are especially significant. For many diseases, particularly those that occur rarely and are poorly understood, identifying genes involved in the disease process is the first step toward improved detection and treatment strategies. Now that a reference set of genes has been identified, scientists can begin to better understand how gene variations are responsible for genetic disease. One type of variation in DNA sequence between individuals is called a single-nucleotide polymorphism, or SNP. We will talk about the role of SNPs in diagnosing genetic diseases later in this booklet.

Although the approximate number of human genes is now known, it will be many years before some of the greatest secrets in the genome will be revealed. As we have already discussed, understanding how genes function will be an enormous challenge. Long after the Human Genome Project, scientists will be challenged to understand how our all of our genes work and how genes interact.

"TAKE HOME" MESSAGE

The past few pages have provided a basic overview of some of the major findings of the genome project. Below is a top-ten list as a summary of important highlights to remember about what we have learned from the Human Genome Project:

1. The human genome consists of ~3.1 billion base pairs; 2.85 billion base pairs have been fully sequenced.
2. The genome is approximately 99.9% the same between individuals of all nationalities and backgrounds.
3. Less than 2% of the genome codes for genes.
4. The vast majority of our DNA is non-protein-coding, and repetitive DNA sequences account for at least 50% of the noncoding DNA.
5. The genome contains approximately 20,000–25,000 protein-coding genes.
6. Many human genes are capable of making more than one protein, allowing human cells to make perhaps 80,000–100,000 proteins from only 20,000–25,000 genes.
7. Functions for over half of all human genes are unknown.
8. Chromosome 1 contains the highest number of genes. The Y chromosome contains the fewest genes.
9. Over 50% of the human genome shows a high degree of sequence similarity to genes in other organisms.
10. Thousands of human disease genes have been identified and mapped to their chromosomal locations.

Learning About the Human Genome Project via the Internet

With emergence of the Internet as an information and communication tool and the almost simultaneous expansion of genetic knowledge from the Human Genome Project, both scientists and laypersons now have easy access to detailed and very current information on the progress being made by the Human Genome Project. There are a number of genome-related sites available on the Web. In this section, I'll describe a few of my favorites. I recommend these student-friendly and easy-to-navigate sites as outstanding resources that provide a wealth of up-to-date information. Use these sites for your own knowledge and enjoyment, as credible resources when writing papers and reports, and for reference when asked by a friend or family member about a genetic disease.

SITES TO VISIT AND INTERNET EXERCISES

I recommend that you begin your Internet search of the genome by visiting the **DOEgenomes** site of the U.S. Department of Energy (DOE) at http://www.doegenomes.org/. This site is an excellent resource that serves as an informative introduction to the Human Genome Project. The "Human Genome Project Information" link (http://www.ornl.gov/sci/techresources/Human_Genome/home.shtml) is *the best* place to start for general information about the history of the project; its goals; answers to frequently asked questions (FAQs); ethical, legal, and social issues; and sequencing and mapping technologies. Quite simply, you will be hard pressed to find a more accurate and informative site on the Web. Figure 12 shows a screen shot of the Human Genome Project Information page indicating major headings and subtopics to a variety of genome links. Can't-miss links include:

- *What is the Human Genome Project?* Takes you to a series of pages that describe the primary goals and objectives of the Human Genome Project with a number of important links to related topics. Provides the history of why DOE became interested in the project, a basic introduction—especially helpful for nonscience majors—on DNA structure and the genetic code, primary goals of the project, a basic explanation of different types of chromosome maps and their importance and applications, a discussion of the role that model organisms have in genetics, a basic overview of DNA sequencing technologies, and discussions on ethical issues surrounding the genome project.
- *Primer: Genomics and Its Impact on Science and Society.* Links to pages that provide a good basic overview on the organization of genetic material in human cells.
- *Science Behind the Human Genome Project: Understanding the Basics.* Presents some of the basic facts about DNA structure, chromosomes, and genes for the newcomer and discusses implementation and goals of the Human Genome Project.
- *Ethical, Legal, and Social Issues.* Exactly what the title says—an excellent resource.
- *Frequently Asked Questions (FAQs).* A great place to begin seeking answers to some of the questions you may have.
- *Facts About Genomic Sequencing.* Student-friendly pages of text, figures, and video clips describe common and new methodologies for DNA sequencing. This site also presents a glossary

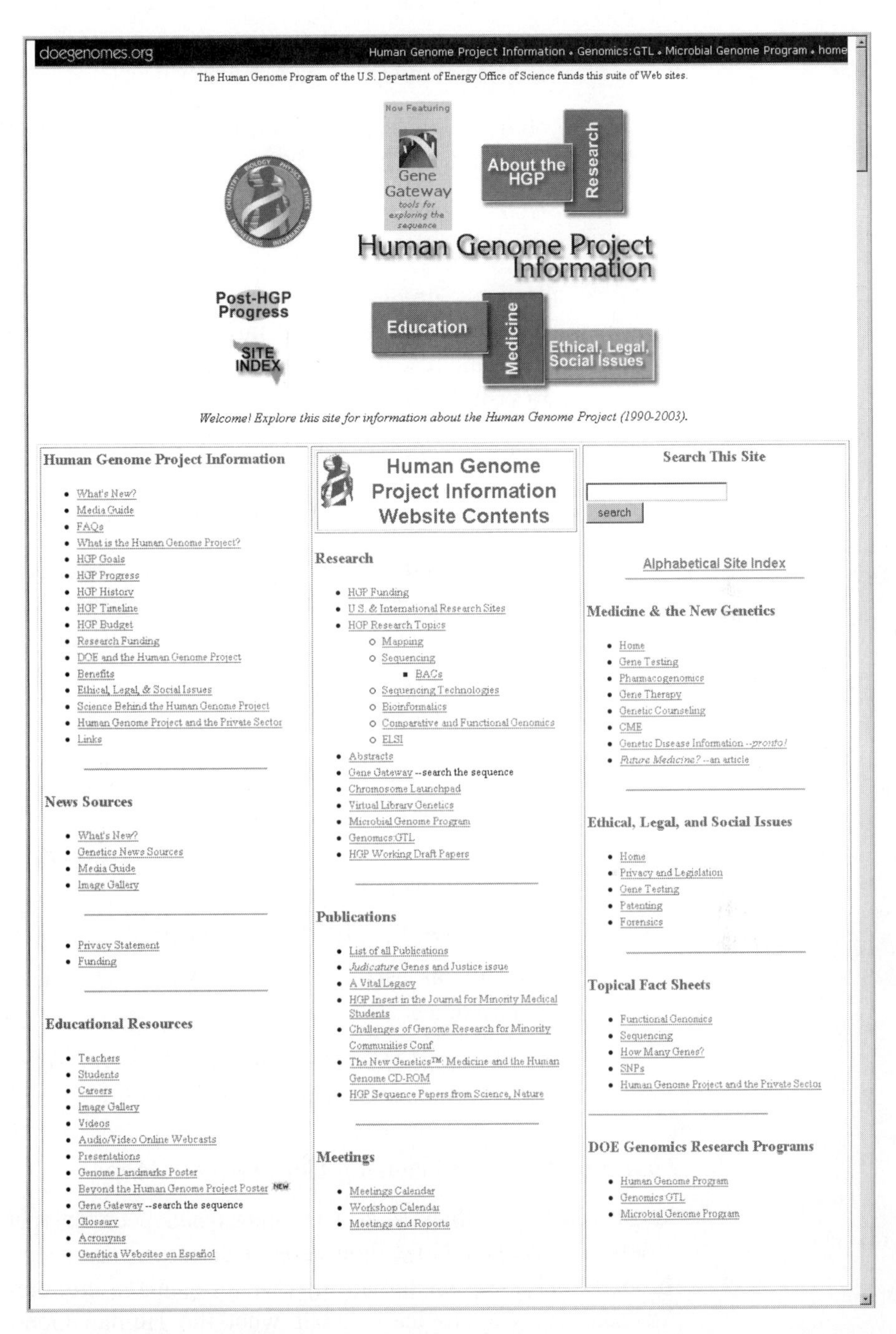

FIGURE 12 Human Genome Project information site from the DOE genomes Web site
Source: U.S. Department of Energy Human Genome Program.
http://www.ornl.gov/sci/techresources/Human_Genome/home.shtml.

of genetic terms and discusses sequencing the genomes of model organisms as part of the project.

- *HGP Progress.* Many informative links to project updates, milestones, individual chromosome data, and much more.
- *HGP Timeline.* Provides an informative timeline with links to relevant articles.
- *Human Chromosome Launchpad.* Links to detailed information on identified genes, gene maps, sequence data, genetic disorders, and research efforts for each human chromosome. Be sure to visit this link!
- *Medicine and the New Genetics.* Links on gene therapy, pharmacogenomics, and topics related to how medical practice in the future will be changed by genetic information.
- *Student Guide to the Human Genome.* These pages incorporate many of the links designed for students that were described above. Another link that shouldn't be missed!

The Howard Hughes Medical Institute has a great site called **Blazing a Genetic Trail** (http://www.hhmi.org/GeneticTrail). There are over a dozen excellent links. In particular, visit the following links:

- *Why So Many Errors in Our DNA?* A nice discussion on how mutations arise and consequences of mutations.
- *Stalking a Lethal Gene.* An informative overview of how the gene for cystic fibrosis was mapped.
- *Reading the Human Blueprint.* A basic overview of genetic technologies.
- *Of Mice and Men.* Excellent discussion on the role of mice as model organisms for studying and curing genetic diseases.

Finding a Gene and Viewing Chromosome Maps

As of 1989, less than 12 disease genes had been mapped by traditional genetic approaches. Now, thousands of disease genes have been discovered through the Human Genome Project and organized into chromosome maps. An excellent way to learn about what the Human Genome Project has revealed is to review some of the chromosome maps available on the Web. For instance, suppose you were interested in the Y chromosome. How could you find out what genes are located on this chromosome? One way to find out why this chromosome is partially responsible for making the strange creatures we call human males is to review maps of the Y chromosome and descriptions of the genes found on this sex chromosome. If you are a male, you can whet your curiosity for genetic clues as to why females are the way they are by learning about genes on the X chromosome!

I encourage you to use the sites presented here to learn about a gene or chromosome of interest. Also, these sites are great resources for finding information on a gene you may have heard or read about but want more information on. If a friend or family member asks you about a gene or a chromosome of interest, you can take them to these sites and uncover more details about that gene than you may ever want to know. I often suggest that students go to these sites to learn more about a rare disease gene related to a disease condition affecting someone close to them. These sites present up-to-date information that can't be found in even the most recent books. If you can't find a gene of interest in these sites, it probably hasn't been identified yet. The sites described here are among the best sites for learning about human disease genes and chromosome maps.

Online Mendelian Inheritance in Man (OMIM) at http://www.ncbi.nlm.nih.gov/entrez/query.fcgi?db=OMIM is a great database of human genes and genetic disorders. In the keyword box, type in the name of a gene or disease you may be interested in. For example, type in "breast cancer," then click the search button. When the next page appears, you will see a list of genes implicated in breast cancer along with corresponding access numbers highlighted in blue as links. Click on one of the links for a gene from this list that you are interested in, and you will be taken to a wealth of information about that gene including background information, links to scientific papers about the gene, gene maps, and even nucleotide and protein sequence data (when available). You might also want to search this site to see if a gene has been found for a particular behavioral condition (for example, alcoholism or depression). OMIM is perhaps the best site to use when searching for information on a gene of interest.

Similarly, the **Weizmann Institute of Science** in Israel (http://bioinfo.weizmann.ac.il/cards/index.shtml) has an excellent site of "gene cards" that provide summaries of information about known genes.

The National Center for Biotechnology Information (NCBI) sponsors **The HumanGenome** Web site (http://www.ncbi.nlm.nih.gov/genome/guide/human/). This site is a nice way to access many of the sites mentioned previously. Click on "Genes & Disease (GD)" or go directly to http://www.ncbi.nlm.nih.gov/disease/ for a student-friendly set of pages on selected disease genes that have been mapped to chromosomes. Clicking on each gene takes you to detailed information about the gene.

Some of the **DOEgenomes** links also provide excellent pages with chromosome maps of identified genes. Visit http://www.ornl.gov/hgmis/posters/

chromosome and http://www.ornl.gov/hgmis/launchpad. See Figure 13 for an example of mapped disease genes on chromosomes 13 and 21 adapted from this site. Notice how genes for several types of cancers are shown on these chromosomes.

Genome Issues

Mike, a physically fit and intelligent college student, walks into the campus pub to order a nonalcoholic drink. Mike is approached by Lauren, an attractive student in his genetics class. Lauren finds Mike interesting. Mike and Lauren exchange small talk, and Mike asks Lauren for a date. While Mike pictures the perfect date, Lauren wonders if she and Mike are genetically compatible. Before giving Mike an answer, Lauren asks to see Mike's "carrier card." Lauren wants to see what genes Mike might be a carrier for before deciding if it's worth her time to go on a date with him. Why waste time dating someone if that person is genetically inferior to you or if there is a chance that your mating in the future would lead to children with genetic defects? Rejected by Lauren, Mike quickly focuses his attention across the room to Elizabeth, a student in his sociology class who he hopes is less informed about genome issues than Lauren!

While I hope the preceding scene never becomes a reality, is it really so far-fetched? As we learn more about human disease genes and develop strategies for the screening and detection of these genes, we will have the ability to visit a physician and gain insight into what genes we have—for better or worse.

From the beginning of the Human Genome Project, the Department of Energy and the National Institutes of Health recognized the importance of informing the general public about genome issues in a practical context to help nonscientists make informed decisions. Significant resources have been committed to examine the ethical, legal, and social issues of genome research.

There are many issues related to the Human Genome Project that will impact all of us in the future. Some of these concerns can be addressed more easily than others. Some will involve personal choices and decisions that individuals will have to make—hopefully with a knowledgeable and informed opinion. Many genome issues will be ballot-box issues in the future.

The range of genome issues is far too great to be completely addressed in this booklet. A few of the most controversial issues are the privacy of genetic information; the moral, ethical, and legal dilemmas posed by genetic technology; and the patenting of genetic information.

How can genetic information be used beyond personal and private decisions? Consider some of these issues:

- As a result of the Human Genome Project, we will have a greater ability to screen, diagnose, prevent, and treat disease

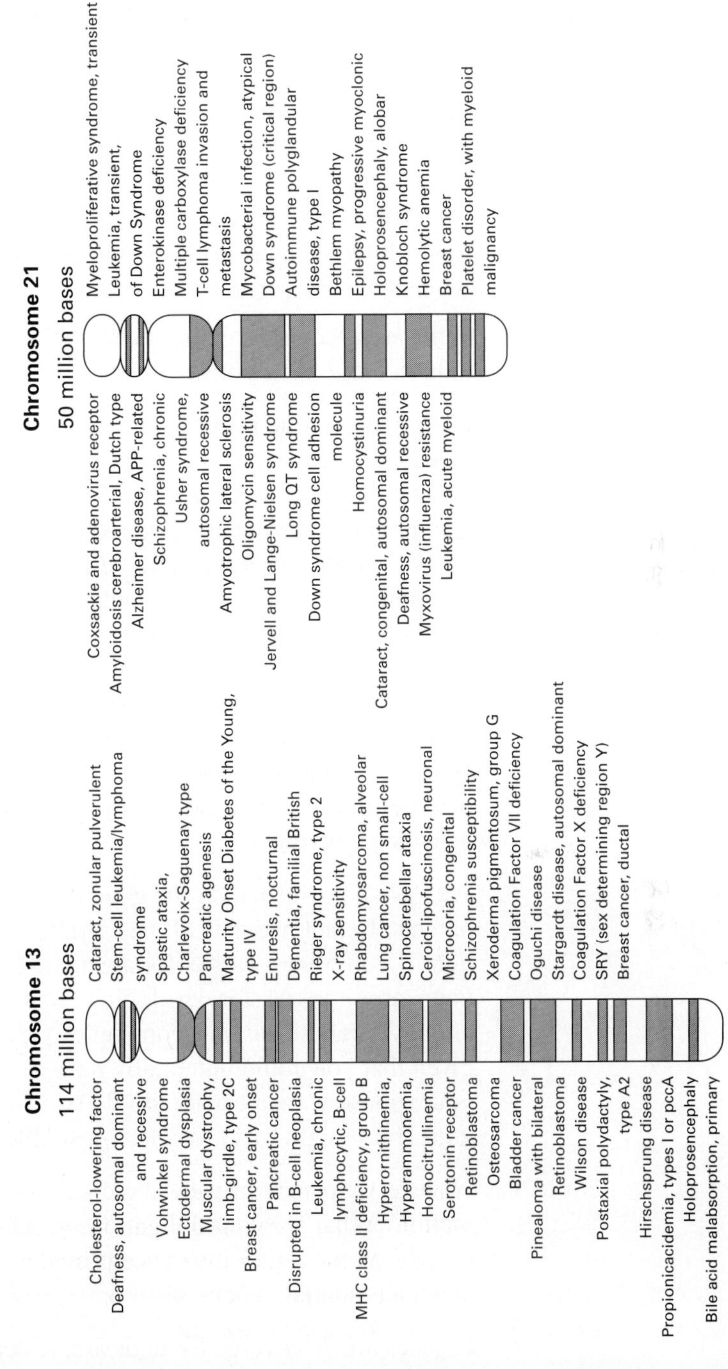

FIGURE 13 Partial maps for disease genes on chromosomes 13 and 21
Source: From: Thieman and Palladino, *Introduction to Biotechnology,* p. 4 Figure 1.2.

conditions. But should we test adults or newborn children for genetic conditions for which there is currently no cure? For example, if one of your parents suffered from a debilitating disease, such as Huntington's disease (a disabling disorder of the nervous system that presents few obvious symptoms until after 30 years of age or so), you have a 50% chance of inheriting the defective gene. Currently there is no cure for Huntington's disease, although a test is available to detect the defective Huntington gene.

- What are acceptable consequences if parents learn that their unborn child has a genetic defect?
- Privacy of genetic information is of considerable concern. Who should have access to your genetic information? Could your genetic background be used to discriminate against you? Should a potential or current employer know about your genetic tests? If so, how might this information affect your job success or ability to obtain and retain a job? Should your insurance company gain access to your genetic information? What if your health or life insurance company learned that you or a family member had a gene that increased your incidence of developing a particular type of cancer? Could they raise your premiums based on "genetic" risk (in the same way that premiums are raised based on other risks, such as how old you are and the car you drive) or deny you coverage?
- We have no control over the genetic lottery we inherit from our parents. In fact, we may all be carriers for some genetic condition. If you were a carrier for a genetic condition, even though you weren't at risk for developing the disease, how might employers and insurance companies view this information, given that you could potentially have a child with the disease (if your spouse were a carrier too)? Days spent at home caring for your child might make you less desirable as an employee.
- How do we ensure privacy and confidentiality of genetic information? What are your obligations to inform others (a potential spouse, employer, or insurance provider) of your knowledge about a potential genetic disorder?
- If genes are discovered for undesirable behaviors, how will these genes be perceived in legal courts if accused criminals use genetics as their basis for a not-guilty, by reason of genetics, plea?

- If you were an employer or director of an insurance company, what access to genetic information would you want, and what would you want to do with this information?

- How would information derived from testing for a genetic mutation affect your life? For many genetic conditions, there is a high likelihood that an individual with a defective gene will enjoy a long, normal, and relatively healthy life span. Think about the dilemmas and stress that would ensue if a genetic test was performed incorrectly, or if inaccurately interpreted or incorrect results leading to a misdiagnosis, for or against a disease, were given to the patient.

- Would society implement mechanisms to prevent or dissuade individuals with genetic defects from having children?

Visit the DOE "Your Genes, Your Choices" Web site (http://www.ornl.gov/hgmis/publicat/genechoice/contents.html) for a thought-provoking series of ethical dilemmas created by genetic testing and genetic technology. What would you do if you had to face the scenarios presented at this site? How would you decide what to do?

Go to Bioethics.net (http://www.bioethics.net) for links on ethical issues related to human genetics and human cloning. Also visit the Web site for The President's Council on Bioethics (http://bioethics.gov). A similar group was originally appointed in 1996 by President Clinton in response to concerns about the potential for human cloning after the cloning of Dolly the sheep was announced. We'll save the pros and cons of organism cloning for another booklet!

Another hot area of debate concerns products of the Human Genome Project. Does a researcher have the right to patent a gene sequence because he or she discovered it first (and wants to claim his or her "property") even if we don't know what it does and there are no clear uses for the DNA sequence? Should scientists be allowed to patent pieces of human DNA? Would patenting slow progress to clone genes if groups hoarded data and didn't share information? Should a group be permitted to stake a claim to a gene, thereby preventing others from working on it or developing a product from it?

Given that in many large labs computers do most of the routine work of genome sequencing, who should get the patent for a sequenced piece of DNA? What about the individuals who figure out *what* to do with the gene? What about patenting a genetically engineered living organism? Bioengineered bacteria (for example, those used to clean up environmental pollution) and genetically altered model organisms have been patented. Can a group claim rights to speculative, anticipated future use of a gene, even if there is no data to substantiate such claims? What if a gene sequence may be involved in a disease process for which a genetic therapy may be developed? What is the best way to use this information to advance medicine and cure disease?

Many scientists believe that it is more appropriate to patent novel technology used to discover and study genes, as well as applications of genetic technology, such as gene therapy approaches, rather than patent gene sequences themselves. What do you think? It is impossible to predict all of the long-term ramifications associated with the Human Genome Project. There are no easy answers to many of these questions. However, knowledge is key to helping us navigate through these issues. Regardless of your major, I encourage you to stay well informed about the power and pitfalls of genetic technology. More than likely, you will encounter some of these issues in your future.

Where Do We Go From Here?

Now that the Human Genome Project is complete, what's next? The legacy of this project will be the wealth of information to be gained from studying the genome in detail. As mentioned earlier, one significant priority of genome scientists will be to learn more about the functions of each human gene—a very challenging task that will take many years of research. Remember, at present the functions for over 50% of human genes remain largely unknown. Many researchers will be working on elucidating the functions of human genes, how genes are regulated, and how different genes affect one another.

Identifying the chromosomal location and sequencing of genes in the human genome will undoubtedly increase our understanding of the complexity of human genetics and the proteins made by human cells. But the substantial task of understanding the functions of all human genes and proteins lies ahead as a major challenge for researchers. Discussion has begun on a possible **Human Proteome Project**, which would focus on the structure and functions of the entire complement of human proteins—a task that will be helped substantially by the identification of the genes that code for all human proteins. Already underway is a 10-year project sponsored by the NIH called the **Protein Structure Initiative (PSI)**. The PSI is designed to examine the three-dimensional structure of human proteins and analyze functions of human protein families based on structural predictions. Data from the PSI will shed light on how proteins functions and provide valuable information that can be used to develop new medicines.

An advanced understanding of human genetic disease will transform medicine as it is currently practiced. However, for this to happen basic research on the functions of human genes and the controlling factors that *regulate* gene expression will provide immeasurable insight into normal gene function and the molecular basis of many human disease conditions. Recently a project called **ENCODE (ENCyclopedia of DNA Elements)** was initiated with the purpose of identifying gene regulatory sequences (many of which reside in noncoding "junk" DNA) that control how genes are turned on or off during transcription.

A whole range of genome studies have been initiated as a direct result of technology advances created by the Human Genome Project. As mentioned earlier, genome projects for many model organisms have been completed and many more are underway. The DOE has established a **Microbial Genome Program** which involves sequencing genomes for a wide range of bacteria, fungi, parasites, protozoa and yeast, including human disease causing microbes such as the Trypanosome *T. brucei* which causes African Sleeping Sickness, the malaria parasite *P. falciparum*, and a wide range of disease causing bacteria. The DOE has also developed a **Genomes to Life Program** which has a number of research goals related to microbial genomes including characterizing microbial genes and proteins and understanding their roles in a wide range of microbial functions, and to exploring novel ways in which microbes can be used to provide energy, clean the environment and carry out other applications.

Human genome pioneer J. Craig Venter left Celera in 2003 to form the J. Craig Venter Institute. One of the Institute's major initiatives is a global expedition to sample marine and terrestrial microorganisms from around the world and to sequence their genomes. Called the **Sorcerer II Expedition**, Venter and his researchers will travel the globe by yacht in a sailing voyage that has been described as a modern day version of Charles Darwin's famous treks on the HMS *Beagle*. A pilot study the Institute conducted on Sargasso Sea off Bermuda yielded around 1,800 new species of microorganisms and over 1.2 million new genes! To date, Venter's research team has detected over 100,000 microbes and sequenced in excess of 4 million genes. This Expedition has great potential for identifying new microbes and genes with novel functions, including commercially valuable genes. For example, the Sargasso Sea project identified thousands of photoreceptor genes. Some microorganisms rely on photoreceptors for capturing light energy to power photosynthesis. Scientists are interested in learning more about photoreceptors to help develop ways in which photosynthesis may be used to produce hydrogen as a fuel source. Medical researchers are also very interested in photoreceptors because in humans and many other species, photoreceptors in the eye are responsible for vision.

Let's consider just a few ways that the Human Genome Project will impact medicine. Genome information has and will continue to result in the rapid, sensitive, and early detection and diagnosis of genetic disease conditions in humans of all ages, from unborn children to the elderly. Increased knowledge about genetic disease conditions will lead to preventive approaches designed to foster healthier lifestyles. Once a disease condition has been identified, new, safer, and more effective approaches for drug therapy and gene therapy will be available.

Every day, new information is helping scientists decipher genetic diseases such as sickle-cell disease, Tay-Sachs disease, cystic fibrosis, cancer, forms of blindness, and certain forms of infertility, to name just a few. Increased knowledge of genes involved in organ development and growth is leading to novel strategies for tissue transplantation and tissue regeneration, including the potential future treatment of spinal cord injuries and neurodegenerative conditions.

From its inception, the Human Genome Project yielded immediate dividends in our ability to identify and diagnose disease conditions. The identification of disease genes has enabled scientists and physicians to screen for a wide range of genetic diseases. This screening ability will continue to grow in the future. Aiding in the diagnosis of genetic disease will be applications involving **single nucleotide polymorphisms** (**SNPs**, pronounced "snips"). SNPs are single nucleotide changes in DNA sequences that vary from individual to individual (see Figure 14). These subtle changes represent one of the most common examples of genetic variation in humans and are largely responsible for the 0.1% differences in DNA sequence between individuals (remember, 99.9% of DNA sequence is the same between all humans).

Most SNPs have no effect on a cell because they occur in non-protein-coding (junk DNA) regions of the genome. But when an SNP occurs in a gene sequence, it can influence traits in a variety of ways, including conferring susceptibility for some types of disease conditions. It has been estimated that SNPs occur every 100 to 300 bases in the human genome. Over 1.4 million SNPs have been identified to date.

SNPs represent variations in DNA sequence that ultimately influence how we respond to stress and disease. Most scientists believe that SNPs will help them identify some of the genes involved in stroke, cancer, heart disease, diabetes, behavioral and emotional illnesses, and a host of other disorders. In 2001, one of the revised goals of the Human Genome Project was to identify SNPs and develop SNP maps of the human genome.

Another relatively new technique for studying genomes will also play an important role in the detection of genetic disease. **Gene microarrays**, also called **gene chips**, are glass microscope slides spotted with genes. Computer-

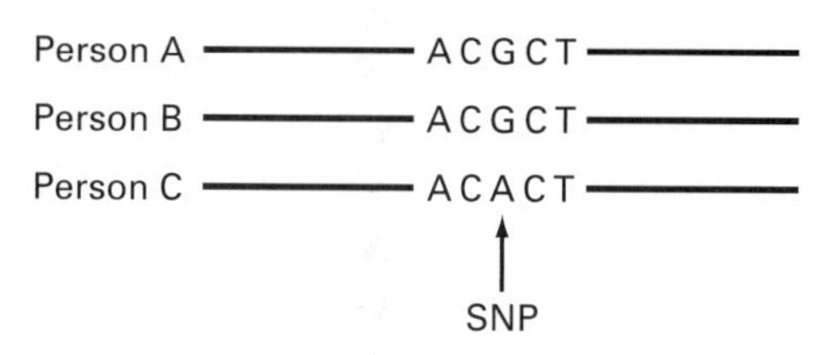

FIGURE 14 Single nucleotide polymorphisms (SNPs)
A small piece of a gene sequence for three different individuals is represented. For simplicity only one strand of a DNA molecule is shown. Notice how person C contains an SNP in this gene. This subtle genetic change may affect how person C responds to a medical drug or influence the likelihood of person C to develop a genetic disease.

automated instruments are used to create tiny spots of DNA on the slide. Each spot contains DNA for one gene. A single microarray can contain thousands of genes. Researchers can use microarrays to screen a patient for a pattern of genes that might be expressed in a particular disease condition. Microarray data can then be used to figure out the patient's risk of developing that disease based on the number of expressed genes for the disease the patient shows.

Researchers are using microarrays in a range of different projects designed to evaluate gene expression in normal cells compared to diseased cells, such as cancer cells, in studies designed to better understand genetic differences that occur in disease states.

Many pharmaceutical and biotechnology companies are heavily involved in the search for SNPs and the development of gene microarrays—good indication of the tremendous potential both approaches may provide for detecting disease in the near future. The discovery of SNPs is partially responsible for the emergence of a field called **pharmacogenomics**. Pharmacogenomics is customized medicine (Figure 15). It involves designing the most effective drug therapy and treatment strategies based on the specific genetic profile of a patient instead of using a "one drug fits all" approach to treating illness.

Consider the following example of pharmacogenomics in action. Breast cancer is a disease that shows familial inheritance for some women. Women with defective copies of the genes called *BRCA1* or *BRCA2* may have an increased risk of developing breast cancer, but there are many other cases of breast cancer in which a clear mode of inheritance is not seen. Perhaps there may be additional genes or nongenetic factors at work in these cases. If a woman has a breast tumor thought to be cancerous, a small piece of cancerous tissue could be used to prepare DNA for SNP and microarray analysis. SNP and microarray data could be used to determine which genes are involved in the form of breast cancer that this particular woman

FIGURE 15 Pharmacogenomics is customized medicine based on a person's genetic profile Source: Spencer Jones/Taxi.

has. Armed with this genetic information, a physician could design a drug treatment strategy—based on the genes involved—that would be *specific* and *most effective* against this woman's type of cancer. A second woman with a different genetic profile for her type of breast cancer might undergo a different treatment.

Many drugs currently used to treat cancer (drug treatment of cancer is known as chemotherapy) may be effective against cancerous cells but also affect normal cells. Hair loss, dry skin, changes in blood cell counts, and nausea are all related to the effects of chemotherapy on normal cells. Wouldn't it be great if drugs could be designed that are effective against cancer cells with no effect on normal cells in other tissues? This may be possible as the genetic basis of cancer is understood and drugs can be designed based on the genetics of different types of cancer. These same principles of pharmacogenomics will also be applied to a wide range of other human diseases.

Based on the pharmacogenomics concepts, some scientists have speculated that **nutrigenomics** will be an emerging field in the future. Nutrigenomics involves developing a personalized diet and exercise plan based on a person's genetic profile. For example, if a person is known to lack genes required for synthesizing vitamin A, that individual would be told to follow a diet plan to supplement vitamin A. Limited applications of nutrigenomics are already being used to screen individuals for genes involved in diabetes and then limiting dietary sugar for these patients, and to screen for genes that elevate blood cholesterol levels to in turn control fat intake.

In addition to advances in drug treatment, **gene therapy** represents one of the ultimate strategies for combating genetic disease. Gene therapy technologies involve replacing or augmenting defective genes with normal copies of a gene. Think about the power of this approach! Currently, there are many barriers that must be overcome before gene therapy becomes a safe, practical, effective, and well-established approach to treating disease. Scientists are working on a variety of ways to deliver healthy genes into humans, but many significant obstacles prevent gene therapy from being widely used in humans. For example, how can normal genes be delivered to virtually all the cells in the body? What are the long-term effects of introducing extra genes into humans? What must be done to be sure the normal protein is properly made once genes are delivered into the body? Great strides toward answering these and other questions are being made using gene therapy in model organisms.

In the future, **stem cells** may be another tool for treating and curing disease. Stem cells are immature cells that can grow and divide to produce different types of cells, such as skin, muscle, liver, kidney, or blood cells. In a lab, stem cells can be coaxed to form almost any tissue of interest, depending on how they are treated. Imagine growing skin and other organs in the lab for tissue and organ transplantation. In the future, it may be possible to collect stem cells from patients with genetic disorders, manipulate these cells by gene therapy, and reinsert them into the patients they were collected from to help treat their genetic disease. Some of this work is already possible, and researchers are working to optimize these technologies.

Disease conditions that result from mutations in a single gene (as in sickle-cell disease) will be the easiest conditions to develop therapies for. Genetic conditions with a multigene basis and multifactorial influences (such as diet, environment, exercise, and stress) will be more difficult to understand and treat because of the complexity of multiple gene interactions. Pharmacogenomics, gene therapy, and stem cell technologies are not the answers to all of our genetic problems, but with continued rapid advances in genetic technology, many seemingly impossible problems may not be so insurmountable in the future.

In some ways, a complete understanding of the human genome may redefine who we are or at least how we perceive who we are with respect to other organisms. The Human Genome Project has confirmed that many human genes share common ancestral origins with genes in other species. Will understanding the genome be the key to unlocking the mysteries of normal biological processes and disease processes, human development, behavior, aging, mental illness, and addictions? In some cases, the answer is likely to be *yes*. But there will be few quick fixes, and only time will tell how great an effect the Human Genome Project will have in these areas.

Completion of the Human Genome Project is not the end of our understanding human genetics but a historic and revealing step that will provide critically important insight about our genome. Current students will become the future generations of scientists and physicians who will advance this work in the years to come. Perhaps as a future biologist, you, too, may contribute to our understanding of the genome and its secrets.

As a biology professor and scientist, I am very excited to see the genome story unfold, and I eagerly anticipate watching powerful applications of the project begin to be realized. I hope you have enjoyed our brief tour of the Human Genome Project and that you have learned something new about this landmark scientific endeavor. I welcome hearing from you about your thoughts about this booklet and the Human Genome Project. Please feel free to contact me at the following address. Good luck with your studies.

Michael A. Palladino, Ph.D.
Associate Professor
Monmouth University
Biology Department
400 Cedar Avenue
West Long Branch, NJ 07764
E-mail: mpalladi@monmouth.edu.
Web: http://bluehawk.monmouth.edu/mpalladi

Resources for Students and Educators

FOR STUDENTS

The following Web sites are among the most informative and student-friendly genomics resources freely available on the Internet. Visit these sites to discover a wealth of information on the Human Genome Project and related topics.

AAAS Functional Genomics Site
(http://www.sciencemag.org/feature/plus.sfg)
Science magazine site provides up-to-date links to a variety of genome topics.

Bioethics.net
(http://www.bioethics.net/)
Site for learning about bioethical issues related to human genetics and cloning.

The Biology Project
(http://www.biology.arizona.edu/molecular_bio/molecular_bio.html)
University of Arizona site provides problem sets and tutorial on DNA structure and gene expression.

Biopharmaceutical Glossaries and Taxonomies
(http://www.genomicglossaries.com)
A good resource with a wide range of links to genome pages and sites that describe some of the language of genomics.

Case Studies in Science
(http://www.ublib.buffalo.edu/libraries/projects/cases/ubcase.htm)
Molecular biology/genetics section presents interesting ethical issues as prenatal genetic diagnosis, sickle-cell anemia, fetal tissue research, and other topics.

Cold Spring Harbor DNA Learning Center
(http://www.dnalc.org)
A student-friendly resource with links to current topics in gene cloning and excellent animations of recombinant DNA techniques.

DNA From the Beginning
(www.dnaftb.org/dnaftb)
Informative overview of the history of DNA and animated primers on the basics of DNA, genes, and heredity.

DOEgenomes.org
(http://doegenomes.org/)
Comprehensive site covering genome programs of the U.S. Department of Energy Office of Science.

DOE Joint Genome Institute
(http://www-hgc.lbl.gov/GenomeHome.html)
Human Genome Sequencing Department home page.

European Bioinformatics Institute (EBI)
(http://www.ensembl.org/)
Provides datasets on eukaryotic genomes, access to chromosome maps, SNP data, and a range of other features.

Federation of American Societies for Experimental Biology Career Opportunity Site
(http://ns1.faseb.org/genetics/gsa/careers/bro-menu.htm)
Tips for careers in genetics and genomics.

Functional Genomics
(http://www.sciencemag.org/feature/plus/sfg/)
Science magazine site provides current headlines in genetics.

GenBank
(http://www.ncbi.nlm.nih.gov/entrez/query.fcgi?db=Nucleotide)
One of the sites most frequently visited by scientists and educators. GenBank is the gene sequence database maintained by the National Institutes of Health. GenBank is the premier resource as a public collection of DNA sequences.

Geneforum.org
(http://www.geneforum.org/)
Designed to inform citizens about advances in genetic research and biotechnology and issues surrounding these sciences.

Genetic Science Learning Center
(http://gslc.genetics.utah.edu)
Interactive, student-friendly animations of DNA structure and replication, transcription, translation, and more.

Genomics Glossary
(http://www.genomicglossaries.com/)
Good resource for keeping up with the vocabulary of genetics.

GenomicsGTL
(http://doegenomestolife.org/)
DOE Genomes to Life Web site.

Howard Hughes Medical Institute: Blazing a Genetic Trail
(http://www.hhmi.org/genetictrail)
Excellent Web site that provides actual stories of gene discovery (such as the search for the cystic fibrosis gene) and dilemmas presented by genetic testing and gene therapy.

Human Chromosome Maps
(http://www.ornl.gov/sci/techresources/Human_Genome/posters/chromosome and **http://www.ornl.gov/sci/techresources/Human_Genome/launchpad)**
DOE sites with current updates on human chromosome maps and gene loci.

Human Genome Program Information Career Page
(http://www.ornl.gov/hgmis/education/careers)
An outstanding site on career possibilities in genetics and genomics, which also includes links to many other valuable resources.

Human Genome Project Information Site
(http://www.ornl.gov/sci/techresources/Human_Genome/home.shtml)
The definitive site for learning about the history, goals, technologies, future directions, and ethical, legal, and social issues of the Human Genome Project. This site also has a wealth of information for educators regarding genome meetings, teaching materials, publications, and medical applications of the Genome Project.

Microbial Genome Program
(http://microbialgenome.org/)
U.S. Department of Energy Microbial Genome Project site.
http://microbialgenome.org/

Model Organisms for Biomedical Research
(http://www.nih.gov/science/models/)
Excellent site for current information on biomedical research using model organisms.

National Center for Biotechnology Information: Human Genome Resources
(NCBI; http://www.ncbi.nlm.nih.gov/genome/guide/human/)
An excellent resource for chromosome maps and disease gene information.

National Center for Genome Resources
(http://www.ncgr.org/)
Good resources for information on genomics and bioinformatics.

National Human Genome Research Institute—National Institutes of Health
(http://www.nhgri.nih.gov)
NIH site on goals and accomplishments of the Human Genome Project.

NOVA Online: "Sequence for Yourself"
(http://www.pbs.org/wgbh/nova/genome/media/sequence.swf)
Outstanding animations on DNA cloning and assembling cloned DNA fragments to sequence segments of a chromosome.

Online Mendelian Inheritance in Man
(http://www.ncbi.nlm.nih.gov/entrez/query.fcgi?db=OMIM)
An outstanding way to search for information on human disease genes.

The President's Council on Bioethics
(http://www.bioethics.gov)
Presidential appointed council charged with reviewing and establishing policy on research involving human subjects.

Science.bio.org
(http://science.bio.org)
Science news Web site provided by the Biotechnology Industry Organization.

Science Odyssey: DNA Workshops
(http://www.pbs.org/wgbh/aso/tryit/dna)
This site provides animations of DNA replication, transcription, and translation.

The SNP Consortium LTD
(http://snp.cshl.org/)
Web site providing a wealth of information on SNPs.

Understanding Gene Testing
(http://newscenter.cancer.gov/sciencebehind/genetesting/genetesting26.htm)
Student-friendly, illustrated guide to gene testing.

University of Pennsylvania Center for Bioethics
(http://bioethics.upenn.edu/)
Home page for the UPenn Center for Bioethics, providing a range of useful and thought-provoking links to topics related to human genetics.

Weizmann Institute of Science (http://bioinfo.weizmann.ac.il/cards/index.shtml)
Good site for information on human genes presented as gene (index) "cards."

Your Genes, Your Choices (http://www.ornl.gov/hgmis/publicat/genechoice/contents.html)
U.S. Department of Energy-funded site explores issues raised by genetic research. Contains excellent case situations and ethical dilemmas for student discussion.

FOR EDUCATORS

Education Resources: Human Genome Program of the U.S. Department of Energy (http://www.ornl.gov/hgmis/education/education.html and **http://www.genome.org.gov/Education/)**
Outstanding sites with a wealth of print and media resources available to educators such as PowerPoint slides (http://www.ornl.gov/sci/techresources/Human_Genome/education/education.shtml)and other materials.

"The Genes We Share with Yeast, Flies, Worms, and Mice: New Clues to Human Health and Disease" (2001). An outstanding report from the Howard Hughes Medical Institute. A limited number of free copies are available to educators. Send a letter on official stationary describing how the publication will be used to: Howard Hughes Medical Institute, Office of Communications; 4000 Jones Bridge Road; Chevy Chase, MD 20815-9864.

The Human Genome Project: Exploring Our Molecular Selves. Limited Edition Multimedia Education Kit produced by the National Human Genome Research Institute of the National Institutes of Health. Designed as an educational tool for high school and college students, the kit contains a commemorative wall poster, a brochure ("Genetics—The Future of Medicine"), a video, and a CD-ROM that contains three-dimensional animations of cells and molecules, timelines in genetics, activities on genome sequencing, a talking glossary, and a discussion of ethical and social issues, among other topics. http://www.genome.gov/Pages/EducationKit/.

Nature's Genome Gateway (www.nature.com/genomics/papers): Free access to published genome research categorized according to organisms. A great collection of genome papers.

Palladino, M. A. (2002). Learning about the Human Genome Project via the Web: Internet resources for biology students. *The American Biology Teacher,* 64, 110–116. Paper describing sites and exercises for teaching majors and non-majors about the Human Genome Project via the Internet.

Primary Research Papers: The following publications are available free online through the journals *Nature* (www.nature.com/genomics) and *Science* (www.scienceonline.org), respectively. These are landmark papers on the Human Genome Project.

IHGS Consortium. (2004). Finishing the euchromatic sequence of the human genome. *Nature,* 431, 931–945.

IHGS Consortium. (2001). Initial sequencing and analysis of the human genome. *Nature,* 409, 860–891.

Venter, J. C., et al. (2001). The sequence of the human genome. *Science,* 291, 1304–1351.

JOURNAL ARTICLES

Bamshad, M. J., & Olson, S. E. (2003). Does race exist?" *Scientific American,* 289, 78–85.

Collins, F. S., Morgan, M., & Patrinos, A. (2004). The Human Genome Project: Lessons from large-scale biology. *Science,* 300, 286–290.

POPULAR BOOKS

Cook-Deegan, R. (2004). *The gene wars: Science, politics, and the human genome.* New York: W. W. Norton & Company. Engaging story about controversies, politics, economics, and the personalities involved in the genome project.

Davies, K. (2001). *Cracking the genome: Inside the race to unlock human DNA.* New York: Free Press. An account of goals, problems, issues and competition surrouding the Human Genome Project.

Dennis, C., & Gallagher, R. (Eds.). (2001). *The human genome.* New York: Nature Publishing Group/Palgrave.

Ridley, M. (2000). *Genome: The autobiography of a species in 23 chapters.* New York: HarperCollins Publishers. A clever and engaging tour of the 23 human chromosomes.

Shreeve, J. (2004). *The genome war: How Craig Venter tried to capture the code of life and save the world.* New York: Knopf. A compelling chronicle of the race to finish the genome project.

Note: All of the Web sites and links presented in this booklet were last accessed and verified for accuracy on January 12, 2005.